Karl Thayer Pomeroy McElroy, Willard Dell Bigelow

Canned vegetables

Karl Thayer Pomeroy McElroy, Willard Dell Bigelow

Canned vegetables

ISBN/EAN: 9783743330276

Manufactured in Europe, USA, Canada, Australia, Japa

Cover: Foto ©berggeist007 / pixelio.de

Manufactured and distributed by brebook publishing software (www.brebook.com)

Karl Thayer Pomeroy McElroy, Willard Dell Bigelow

Canned vegetables

U. S. DEPARTMENT OF AGRICULTURE.

DIVISION OF CHEMISTRY.

BULLETIN   No. 13.

FOODS

AND

FOOD ADULTERANTS.

INVESTIGATIONS MADE UNDER DIRECTION OF

H. W. WILEY,

CHIEF OF THE CHEMICAL DIVISION.

PART EIGHTH.

CANNED VEGETABLES.

BY

K. P. McELROY, Second Assistant Chemist,

WITH THE COLLABORATION OF W. D. BIGELOW.

PUBLISHED BY AUTHORITY OF THE SECRETARY OF AGRICULTURE.

WASHINGTON:
GOVERNMENT PRINTING OFFICE.
1893.

LETTER OF TRANSMITTAL.

U. S. DEPARTMENT OF AGRICULTURE,
DIVISION OF CHEMISTRY,
Washington, D. C., June 7, 1893.

SIR: I have the honor to submit herewith for your inspection and approval the manuscript of Part Eighth of Bulletin No. 13, on canned vegetables.

Respectfully,

H. W. WILEY,
Chemist.

Hon. J. STERLING MORTON,
Secretary of Agriculture.

CONTENTS.

	Page.
Letter of transmittal	III
Introductory and summary. By H. W. Wiley	1015
Scope of the work	1015
Methods of preserving	1015
Preservatives employed	1016
The use of copper and zinc	1016
Vessels used	1018
Food value and digestibility of canned goods	1020
General remarks	1021
Report of investigations and analyses. By K. P. McElroy and W. D. Bigelow.	1022
Historical notes	1022
Scope of the investigation	1027
Methods for proximate analysis	1027
General examination	1027
Water	1028
Ether extract	1028
Crude fiber	1028
Albuminoids	1028
Digestible albuminoids	1028
Ash	1029
Salt	1029
"Corrected ash"	1029
Preservatives	1029
Method for detection of preservatives	1030
Sulphurous acid	1032
Salicylic acid	1033
Saccharin	1034
Benzoic acid	1035
Hydronaphthol	1035
Boric acid	1035
Tin plate in cans	1035
Use of solder in canning	1037
Lead-topped bottles	1038
Metallic contaminations	1039
Lead	1039
Zinc	1041
Tin	1041
Greening vegetables with salts of copper	1042
Copper-greening in France	1046
Copper-greening in Belgium	1066
Copper-greening in Germany	1066

CONTENTS.

	Page.
Greening vegetables with salts of copper—Continued.	
Copper-greening in Italy	1072
Copper-greening in Great Britain	1073
Copper-greening in the United States	1074
Analytical data	1074
Samples bought	1074
Peas	1075
"Haricots verts"	1097
String beans	1099
Stringless beans	1103
Haricots flageolets	1106
Haricot panachés	1107
Little green beans	1108
Wax beans	1109
Lima beans	1109
Baked beans	1113
Red kidney beans	1117
Corn	1118
Artichoke	1127
Sweet potato	1129
Okra	1129
Brussels sprouts	1131
Tomatoes	1131
Asparagus	1134
Pumpkin	1138
Squash	1139
Macédoine	1139
Succotash	1141
Mixed corn and tomatoes	1144
Mixed okra and tomatoes	1145
List of packers whose goods were examined	1146
Appendix	1159
Prohibition of sale of coppered pickles in Brooklyn, N. Y	1159
Sale of canned vegetables colored with salts of copper	1159
Addition of sulphate of copper to canned vegetables	1161
Imported canned goods	1161
Apparatus for cooking vegetables and fruits	1162
Copper sulphate in green peas	1162
Presence of metallic compounds in alimentary substances	1162
Note on copper in vegetables	1162
Copper in preserved green peas	1163
Analysis of canned peas	1163
Rubber rings in the preserving industry	1163
Metal vessels for culinary purposes	1163
On the occurrence of tin in articles of food and drink and the physiological action of tin compounds	1164
Use of tin cans for preserving	1165
Poisonous action of tin	1165
Technical determination of zinc	1166
Detection of benzoic acid	1167

FOODS AND FOOD ADULTERANTS.

PART VIII.—CANNED VEGETABLES.

INTRODUCTION AND SUMMARY.

By H. W. Wiley.

SCOPE OF THE WORK.

In undertaking the study of canned and preserved foods it was the original intention to have the whole of the work included in one publication. After the inception of the work, however, its magnitude was found to be so great as to render the completion of the original plan impracticable. It has therefore been thought best, without waiting to finish the whole of the work on canned goods, to prepare for publication that part of it relating especially to canned vegetables.

The work with canned vegetables has been directed especially to the methods of preserving, the preservatives employed, the character of the vessels in which the goods are preserved, and to their food value and digestibility.

METHODS OF PRESERVING.

A brief history is given of the process of preserving foods by sealing them at a high temperature from contact with the external air. It is shown that it was originally believed that the success of this process was due to the exclusion of the oxygen. The error of this, however, is set forth and the true theory developed, which rests upon the fact that the germs or microörganisms capable of inducing decay of the food are killed by the high temperature. The exclusion of the external air prevents the access of new germs, and thus the foods are preserved simply because the organisms which produce putrefaction can not be introduced.

It is shown that a temperature high enough and sufficiently prolonged to kill these germs in vegetables tends to disintegrate many of them and render them less attractive to the eye than when in the natural state. For this reason canners have sought other methods of preserving the foods in such a way as not only to preserve them from decay, but also to preserve their natural attractiveness. The preservatives which have been used for this purpose, and which have been found to the largest extent, are salicylic acid and sulphurous acid, the latter usually in the form of sulphites. Other preservatives are also sometimes used, such as boric acid, saccharin, etc. The action of all these added preservatives, together with a discussion of their physiological effects, as gathered from the experience of physicians and others, forms a prominent part of the bulletin.

PRESERVATIVES EMPLOYED.

Opinions are divided in regard to the wholesomeness or unwholesomeness of these added preservatives, the great weight of testimony being to the effect that while these bodies in small quantities are not injurious to health, yet the continual use of them, even in such small quantities, may finally become prejudicial. It is also shown that the same qualities which enable these preservatives to prevent the action of microörganisms, and thus preserve the food from decay, are also active in the digestive organs and hinder the normal functions of the digestive ferments. In other words, the forces which tend to preserve in this way the vegetables from decay also tend in like manner to retard the processes of digestion. The fair conclusion from the data which follow in this bulletin is, without doubt, that the use of added preservatives in canned vegetables is objectionable. This conviction, however, is not strong enough to warrant the absolute inhibition of these bodies, but the consumer would be sufficiently protected if the law should require that on each can of preserved vegetables a statement should be found as to the character of the preservative used and the amount of it which has been added. The consumer and his medical adviser are thus properly forewarned of the danger which they may encounter in the way of such foods, and if in the face of this announcement they see fit to continue their use, it is a matter which rests solely with them and they can not hold the guardians of the public health responsible for any ill effects which may follow. Concisely, the views which we have reached as a result of these investigations are these: First, that the use of added preservatives is, upon the whole, objectionable; second, that their absolute inhibition is not warranted by the facts which have come to our knowledge, but in all cases their presence should be marked upon the label of the can.

THE USE OF COPPER AND ZINC.

There are other added chemicals which are found in many varieties of canned vegetables, which are used not especially for the purpose of preserving them, but for adding to the attractiveness of their appearance. I refer chiefly to the use of copper and zinc salts to secure and preserve the green color of canned peas, beans, etc. The use of copper for this purpose is a very old one. Long ago it was observed that the cooking of peas, beans, and other green vegetables in imperfectly cleaned copper vessels would secure a deeper and more attractive green appearance for the cooked product. It did not take the observing cook long to discover that this improvement in appearance was due to the copper or zinc present in the copper or brass vessels. The same effect was found to be produced when these vegetables were cooked in ordinary vessels with the use of small quantities of copper or zinc salts. Upon the whole, copper salts were found more convenient for this purpose, and hence at the present day an immense industry has grown up in the greening of canned vegetables by the use of copper and zinc, especially of the former. By consulting the analytical data which follow, it will be found that a large part of such canned goods exposed for sale in this country has been greened by the addition of copper, and in some cases of zinc. For instance, the amount of copper found in peas of French origin was uniformly much greater than that found in American canned peas. Of forty-three samples of American canned peas examined 32.56 per cent were found to contain no copper, while 67.44 per cent were colored with copper. Of thirty-six samples of French peas all were colored with copper, except one, which was colored with zinc. In regard to the quantity of copper found the following comparison will be of interest:

Amount of copper per kilogram.	American peas.	French peas.
	Per cent.	Per cent.
Less than 10 mg copper per kilo	30.23	0.00
Between 10 and 18 mg copper per kilo	11.63	5.74
Over 18 mg copper per kilo	25.58	94.29
Over 25 mg copper per kilo	16.28	88.57
Over 50 mg copper per kilo	6.98	60.00
Over 75 mg copper per kilo	0.00	31.43
Over 100 mg copper per kilo	0.00	11.43

The literature on this subject, it will be found, has been carefully collated in the pages which follow, and, as in the case of added preservatives, it is difficult to come to a definite conclusion in the matter. Almost the same statements may be made in regard to the use of greening materials as have been made in respect of added preservatives.

The occasional use of a small quantity of a copper or zinc salt, it must be allowed, can be practiced without practical injury to health. On the other hand, the continual and regular consumption of even the small quantities of these materials present in canned vegetables must be regarded as at least prejudicial to health. Therefore it is concluded that the public health will be sufficiently conserved provided each can of vegetables which has been greened artificially in this way shall bear plainly marked upon the label the nature of the greening material and the amount thereof employed. The responsibility of the use of these vegetables will then be thrown upon the consumer and he can exercise his own judgment in regard to the matter.

The question of the use of copper in canned goods has been agitated in France for nearly a quarter of a century. At first the committees appointed by the Government to investigate the matter reported uniformly against the use of copper for greening. While French packers were not allowed for some time to sell their copper-treated goods to French consumers, they were not prevented from using copper when the goods were intended for export. For instance, in 1875 some Bordeaux packers labeled their goods "Green peas (or beans) greened with sulphate of copper. Made specially for exportation to America and England, and not sold for French use." Copper was present in some of these samples to the extent of 40 mg per kilo. After this practice had gone on for some time the board of hygiene of the Gironde concluded to prohibit it, stating that no distinction should be made between goods destined for exportation and those intended for home consumption. Nevertheless, there was such a demand for goods of this kind that the exigencies of commerce gradually got the better of the hygienist, with the result that the French Government has finally permitted the use of copper in greening canned vegetables, requiring, however, that some definite mark shall be used in connection therewith. The canners, however, were shrewd enough to elude the necessity of marking their goods as having been greened with copper or zinc and fulfill the letter of the law, if not the spirit, by marking them with some indefinite mark such as *à l'anglaise*. The result is that the purchaser of these goods has no intimation, as far as the label is concerned, of the nature of the material which is employed in greening, and the canners themselves claim that if they were compelled to mark their goods as having been greened with copper or zinc it would entirely destroy their sale. The question here is one of sight and not of taste or digestive value, and it seems that it would be wise to recommend to the consumer of canned goods to be content to use them, even if they are slightly pale or yellow, rather than to have them of a bright green color at the possible expense of health and comfort. The vast extent of the practice of greening foods, together with the amounts of greening material which have been found in the different cans, will be seen by consulting the analytical details which follow.

VESSELS USED.

Another prominent feature of the work which we have conducted is found in the examination of the vessels containing the vegetables. In Germany the law requires that the tins employed for holding the canned goods shall not contain more than 1 per cent of lead. In this country there is no restriction whatever in regard to the character of the tin employed, and as a result of this the tin of some of the cans has been found to contain as high as 12 per cent of lead. There is no question whatever among physiologists in regard to the effect of lead salts upon the human system. The continual ingestion of even minute

LEAD IN CANNED VEGETABLES.   1019

quantities of lead into the system is followed eventually by the most serious results. Painters' colic, lead palsy, and other serious diseases well known to physicians, are the direct effects of the continual exposure of the system to the influence of minute portions of lead salts. Therefore the greatest care should be exercised in the preparation of canned foods to exclude every possibility of the ingestion of lead. Even tin salts are poisonous, but not to the extent of lead, so that the presence of a minute portion of tin in canned vegetables, coming from the erosion of the cans containing them, is not a matter of such serious import as the presence of lead. Perhaps it would be quite impossible to exclude tin absolutely from canned goods when they are canned in tin, but it is possible to exclude the salts of lead. This can be done by requiring that the tin shall not contain more than, say, 1½ per cent of lead, and, in the second place, that the solder which is employed shall be as free from lead as possible. In Germany the solder employed in sealing the cans is not allowed to contain over 10 per cent of lead, while in this country the analyses of numerous samples of the solder employed show that it contains fully 50 per cent of lead. In addition to this there is no care taken to prevent the solder from coming in contact with the contents of the can. It is a rare thing to carefully examine the contents of a can without finding pellets of solder somewhere therein. Often on examining the inside of a can it is found that large surfaces of solder on the seams are exposed to the action of the acid contents. The result of all this is, as will be found by consulting the analytical data which follow, that lead is a very common constituent of canned goods.

Another great source of danger from lead has been disclosed by the analytical work, viz, in the use of glass vessels closed with lead tops or with rubber pads in which sulphate of lead is found to exist. As a sample of this may be mentioned the goods of Eugène Du Raix, of Bordeaux. All the samples of his goods examined were put up in lead-topped glass bottles. All except one contained salicylic acid and all of them save one contained copper. In one of these samples lead existed to the enormous amount of 35.2 mg per kilo; in another 15.6 mg per kilo were found, while in one sample, viz, No. 10937, the extraordinary quantity of 46 mg per kilo was discovered.

It is not difficult to see how goods covered with lead tops can be contaminated. It may be claimed that these goods should never be turned upside down, but the shippers pay little attention to such directions and the result is that the goods may be kept for days or even weeks in such a position as to bring the contents of the can in contact with the lead tops or with the rubber pads containing lead. The constant consumer of such goods, therefore, must run some risk of being exposed to the insidious inroads of some of the diseases peculiar to the action of small quantities of lead upon the human organism. It is not too much to ask that the law should require the canners to exercise the utmost care to exclude all dangers of this kind.

The general result of the examination of the canned goods exposed for sale in this country leads to the rather unpleasant conclusion that the consumers thereof are exposed to greater or less dangers from poisoning from copper, zinc, tin, and lead. These dangers could be easily removed if the manufacturers of these goods were required to follow the dicta of a reasonable regard to public hygiene.

FOOD VALUE AND DIGESTIBILITY OF CANNED GOODS.

In regard to the food value of canned goods, interesting data have also been obtained. It will be seen that many expensive articles of canned goods contain an amount of nutrient matter totally out of proportion to the price paid therefor. The conclusion is therefore forced upon us that the use of canned goods is in every sense a luxury and a luxury which is attended with many dangers. On the whole, the less rich portions of our population should rather congratulate themselves that their incomes do not warrant them in purchasing at a high price foods of so little digestive value and fraught with so many dangers to health. As an illustration of the excessive cost of some goods put up in cans, attention may be called to the analytical data in the tables which follow. These tables will be useful to consumers who have not time to search through all the details of the bulletin.

The quantity of dry food material in canned goods varies within wide limits. It is very low in such vegetables as string beans, asparagus, etc., and quite high in such materials as canned corn, succotash, and other bodies of that description. The lowest percentage of dry matter in string beans of American origin was 4.17. In other words, in buying 100 pounds of such material the consumer purchases 95.83 pounds of water.

The price of the packages of string beans varied within wide limits, depending both upon the size of the packages and the labels they bore. The highest price paid was 35 cents, and the weight of the contents of the package was a little over 3 pounds. The lowest price paid was 10 cents, and this was paid in many instances. The highest price paid, according to the percentage of dry matter, was in sample 10928, costing 30 cents and containing only 254 grams of string beans, 31.1 grams of dry matter, and 94.37 per cent of water. The price of the dry matter in this package was nearly 1 cent per gram, which would be almost $5 per pound. The enormous cost of food in canned goods is illustrated to the fullest extent by this sample, showing in a striking way that such food materials must be regarded in the light of luxuries or condiments rather than as nutrients to support a healthy organism.

In regard to the composition of the dry material of string beans of American origin, full data are found in the analytical tables in the body of the report. To illustrate its nutrient value it may be well to give the analysis of the sample just mentioned, viz, 10928. The dry matter of this sample contained 0.46 per cent of matter soluble in ether, presumably of an oily or fatty nature; 8.67 per cent of indigestible fiber; 25.5 of mineral matter, of which 18.37 per cent was common salt and 6.68 per cent of other mineral substances. Of nitrogenous matter in the form of albuminoids it contained 16.16 per cent, of which 11.23 per cent were digestible. Of carbohydrates, including sugar, starch, etc., it contained 49.63 per cent. Of the total solid matter present only 69.19 per cent were digestible. We have here a substance which cost nearly $5 per pound, and of which, in round numbers, only 70 per cent were digestible. Thus the digestible matter cost about one-third more, or about $6.50 per pound.

In regard to the use of common salt in these canned vegetables, it may be said that as a rule it is added as a condiment and not as a preservative. The proportion of it in relation to the whole contents of the can is not very high, but the percentage in the dry matter of the can is very considerable. In one instance, viz, 10923, of American string beans, it was found that 40.58 per cent of the dry matter consisted of salt. In this case the salt evidently had been added either as a preservative or with the fraudulent intent of increasing the weight, more likely as a preservative. The extent to which common salt may be added is a matter which has, I believe, not been regulated by law in any country. There should, however, be a limit even to the addition of this comparatively harmless substance.

The percentage of water in the French haricots verts was even higher than in American string beans. In one instance, No. 10939, the percentage of water found was 96.13.

The percentage of salt in the dry matter of the French product is quite uniform, the maximum being 19.13 per cent and the minimum 8.34 per cent. The percentage of albuminoids is somewhat higher than in American goods, but the digestible albuminoids are in no greater abundance.

GENERAL REMARKS.

A general view of the digestive experiments must lead to the conviction that the process of canning, especially when preservatives are employed, such as salicylic acid and sulphites, tends to diminish the digestibility of the albuminoid and other bodies. The low percentage of digestible albuminoids will be remarked with some degree of astonishment in all the analytical tables.

A careful perusal of the data in the body of the report will not fail to convince every unbiased person that the use of canned vegetables is upon the whole an expensive luxury. It is not the purpose of this investigation to discourage the use of such bodies, but only to secure to the consumer as pure an article as possible. Nevertheless these practical conclusions may prove of some help to the laboring man and the head of a family, when he finds himself in straightened circumstances, by assisting him in investing his money in a wiser and more economic way than in the purchase of canned vegetables. An expenditure of 10 or 15 cents for a good article of flour or meal will procure as much nutriment for a family as the investment of $3 or $4 in canned goods would.

The investigations which are recorded in the accompanying report were made upon the following canned vegetables: Artichokes, asparagus, beans, Brussels sprouts, corn, okra, peas, pumpkin, squash, sweet potatoes, tomatoes, macédoine, mixed corn and tomatoes, mixed okra and tomatoes, and succotash.

The samples were purchased in the open market in Washington, D. C., Schuyler, Nebr., Kissimee and Orlando, Fla. Dealers were not acquainted with the purpose of the purchase, and it is believed that the goods represent fairly the character of the canned vegetables found in the markets of the United States.

The analytical work was conducted by Messrs. K. P. McElroy and W. D. Bigelow, assisted by Messrs. T. C. Trescot, Gus. Wedderburn, and E. G. Runyan. Mrs. K. P. McElroy voluntarily contributed largely to the successful issue of the investigations. The work has been one of great magnitude and has consumed more time than was originally intended for this purpose. The character of the work, however, and the value of the data which have been secured fully compensate for the expenditure of the additional time required to complete this branch of the investigation.

REPORT OF INVESTIGATIONS AND ANALYSES.

By K. P. McElroy and W. D. Bigelow.

HISTORICAL NOTES.

The process of preserving food by canning in its present form appears to date back to the patent of Pierre Antoine Angilbert in 1823, though it is said to have been in use at least three years earlier.[1] The method described by Angilbert does not vary essentially from the present practice. The food, together with some water, was placed in a tin can, a lid carrying a minute aperture fastened on and heat applied. When the liquid in the can boiled briskly and all air was expelled, the hole was closed with a drop of solder.

Preserving food in bottles instead of cans, but by a method identical in principle with that just described, is an older invention. The first record of it appears in a paper submitted to the English Society of Arts in 1807, under the title "A method of preserving fruits without sugar for house or sea stores" by Mr. Saddington.[2] The method there described is to fill bottles with the fruit, loosely cork, place them in a vessel containing cold water, which should reach their necks, and grad-

---

[1] Letheby, Chem. News, (Amer. Repr.) 1869, 4, 74.
[2] Hassall: Food and its Adulterations, London, 1855, 432.

ually raise the heat to a temperature between 70° and 77°, keeping it there for half an hour. The caution was given not to heat higher or longer, as the fruit would be liable to burst. Lastly, the bottles were to be filled with boiling water to within an inch of the neck, corked immediately and laid upon their sides in order that the hot water might swell the corks. After covering the corks with cement, the operation was complete.

The credit for the discovery of this process, however, is generally given to M. Appert, who was the first to put it into use on the large scale. In 1810 M. Appert published a book giving directions for this method of food preservation, for which he was awarded the prize of 12,000 francs offered in 1809 by the French Government for a process of preserving perishable alimentary substances. Appert's directions were to partially cook the food and put it into strong glass bottles, filling them almost to the top. These bottles were then to be securely corked and exposed for some time to the action of boiling water. To guard against accident, each was to be separately covered with a cloth bag, and the water in which they were plunged was to be gradually heated, starting cold. The boiling temperature was maintained for an hour, and then the fire was drawn and the bath and bottles allowed to cool slowly. Peas and beans were to be put in the bottles in the raw state, and the boiling temperature maintained for about two hours.

This method is, of course, founded on the same principle as the bacteriological operation of sterilizing. None of the common bacteria can survive the action of boiling water for more than a few minutes, and, although their spores are more resistent, even these will not usually survive more than half an hour. Moreover, few of the common putrefactive bacteria are spore-producing. The bactericidal action of heat in the canning process is much facilitated by the fact that in most cases the liquids surrounding the canned foods are weakly acid. Bacteria are much more susceptible to heat in an acid liquid than in one which is neutral or alkaline.[1]

For many years the fact of the preservation of foods treated by this process was ascribed solely to the fact that all air is expelled from the can during the process of canning, it being supposed that air was absolutely necessary for the putrefactive process, or at least was necessary to initiate it. This is, of course, not so. Some of the common bacteria causing putrefaction are absolutely anaërobic, and the presence of air is fatal or detrimental to them, so that they flourish in the interior of decaying masses, but are not found on the surface; to others, air is not necessary though not detrimental, while many again flourish only when it is present. It follows that the mere presence or absence of air in the interior of the can is a matter of no importance in itself. Tyndall demonstrated the fact that air played no important part in putrefaction,[2]

---

[1] Tyndall: Floating Matter of the Air (New York: 1882), 203.
[2] *Ibid*, 85 and 88.

save as a carrier of living organisms. Boiled infusions of a great many of the common food materials were subjected by him to the action of air from which solid matter either had been removed or had been rendered inert by heat, and these decoctions did not decay, but remained unchanged for long periods. In this connection he demonstrated the fact that air passed through loosely packed cotton or through a red-hot tube permanently lost its power of infection.

Based on the erroneous idea of the necessity of oxygen to the decomposition of organic materials, many processes were proposed in the first half of the century for the preservation of food by exclusion of air. In 1810 Augustus de Heine proposed to exhaust the vessel containing food by means of an air pump, but the process did not answer. In 1828 Donald Currie proposed an improvement, which consisted in allowing carbonic acid gas to enter after the extraction of the air. Later, Leignette (1836), Bevau (1842), Rettie (1846), and Ryan (1846) proposed other improvements, but none of these were successful.[1] Jones and Trevithick devised an improvement which consisted in admitting pure nitrogen after exhausting the air, once more exhausting, and finally admitting a mixture of nitrogen and sulphurous acid. This process was quite successful, but the preservation of the food was quite evidently due to the antiseptic action of the sulphurous acid and not to the exclusion of air. At present this method has merely a historical interest.

Since the days of Saddington and Appert, preservation of food by their process has become one of the world's great industries. Canneries dot every country of the earth and their product is found on every table. All manner of food is canned, and that at prices which place it within the reach of the humblest pockets. Preserved food has been a great democratic factor, and has nearly obliterated one of the old lines of demarcation between the poor and the wealthy. Vegetables out of season are no longer a luxury of the rich. The logger may to-day have a greater variety of food than could Queen Elizabeth have enjoyed with all the resources and wealth of England at her command. In the American grocery—pineapples from Singapore, salmon from British Columbia, fruit from California, peas from France, okra from Louisiana, sweet corn from New York, string beans from Scotland, mutton from Australia, sardines from Italy, stand side by side on the shelves.

In the United States the canning trade has kept full pace with the wonderful development of the country. Its extent may be judged by taking the statistics for one single product—green corn. In 1892 the pack reached the enormous total of 84,700,000 cans.[2]

Yet great as is this industry it has attracted little attention from those charged to guard the welfare of the people against either the skill or the carelessness of the food adulterator. Abroad a little scat-

---

[1] Letheby, Chem. News (Amer. Repr.), 1869, 4, 74.
[2] American Grocer.

CONTAMINATIONS OF CANNED VEGETABLES.

tered work has been done on canned foods, though nothing systematic, and rich as is the literature relating to food adulterations, singularly little of it has to do with the examination of these goods.

In this country little attention has been paid to the matter. Massachusetts, working under an efficient food and drug law, has done some work toward preventing the sale of sophisticated imported canned foods, but as far as the records show has done little or nothing with American goods.

The Brooklyn board of health for some years has been devoting more or less attention to canned foods, and in particular to those which are coppered.

Canned vegetables are not much subject to adulteration in the restricted sense of the word, which implies the addition of foreign substances to food for the purpose of increasing its quantity. The only practice in vogue which can properly come under this head is the addition of undue amounts of water during the canning process. This often occurs. Additions of salt might be regarded in this way, but this substance is added primarily as a condiment. Of adulteration in the more modern sense, that which includes sophistication, there is a great deal, and indeed it may be said to be almost universal. There are few canners who do not use salicylic acid or other preservatives, and the trade in coppered vegetables has grown to enormous proportions. Besides these wilful additions there is a class of what may be called unintentional sophistications, such as the presence of lead, tin, or zinc in these foods. These substances are often present, but are never, except occasionally in the case of zinc, added intentionally.

Ptomaines are often said to be present in canned foods, and this may sometimes be the case, but their occurrence in canned vegetables must be extremely rare. Ptomaines are by definition the result of bacterial action, and where this action does not occur they must of necessity be absent. Vegetables are usually canned in the fresh state, and if they are in any degree spoiled at the time, the fact is usually conspicuously evident to the taste, so that the canner can not afford to use them. Bacterial action seldom occurs in the can without bursting it or rendering it unsalable. Ptomaines may, however, develop where the canned food is allowed to stand for some time after opening, though even then this is unlikely in the case of preserved vegetables.

It may be said, therefore, that the principal risks to health which may arise from the use of canned goods are those due to the use of preservatives, or to the presence of the heavy metals, copper, tin, lead, and zinc. Iron, though often taken up by the food in considerable quantities from badly tinned cans, may be disregarded in this case, since it is not only a normal constituent of food, though hardly in the forms which it assumes in canned goods, but is not poisonous. Its desirability as an addition to food may be questionable, but it can not be called materially deleterious. In regard to the other substances mentioned, the case is different.

23368—No. 13——2

Lead is extremely poisonous, and tin is also poisonous, though in a much less degree. As to the preservatives in common use, of which salicylic acid may be taken as a type, and the salts of copper and zinc, their toxic action is not yet definitely known. This much is certain, however, that they have a marked physiological action and are all of them more or less potent medically. In large quantities they create very evident symptoms of poisoning, though this is usually only temporary. In the quantities in which they are liable to occur in canned foods, their action is at the best uncertain. They may be innocuous—they may not be. Much evidence can be collected to prove either side of the question. It is a question which science is not yet prepared to settle. Pending that settlement, however, it may be said that their use is to be reprobated, inasmuch as any benefit which may be derived by the canner from their presence he can secure in other and less dubious ways. At the very least any food which contains them should be clearly and distinctly labeled, with the fact expressed in direct language. Where this is not done, their presence should be considered to be an adulteration and punished as such.

If there is any fact which is clearer than another, it is that no man or set of men has any right to administer surreptitiously to any other man a more or less potent drug. Every man has a right to knowledge of the fact of being drugged, unless he expressly waives this right in favor of a physician. Even here the law steps in and prescribes that this physician shall be a member of a recognized school. This the canners seldom are. Salicylic acid, which may be taken as a type of these additions, for instance, is a valued medicine in many cases, is in fact one of the best known remedies for rheumatism, and is believed never to have caused death in any dose.[1] But this is no justification for its use. It is certain that it disturbs the normal course of the bodily functions—it must of necessity do so to have medical value—and this fact is alone enough to demand its exclusion from any food intended for general use, unless the food be so labeled.

There is another thing which may be said on this point. Were it as harmless as distilled water, there would be no excuse for its addition to food without notification to the consumer. Salicylic acid is not a normal constituent of any common food, and its addition to such foods for any purpose and in any quantity, without due notice to the consumer, is plainly adulteration. If any man desires to have salicylic acid in his food there is no doubt of his right to have it, since it is not a sufficiently violent poison to warrant the Government's forbidding him. But there is also no doubt of the fact that the canner has no right to admix it

---

[1] There are several cases on record of death supposed to have been due to this substance, notably the one reported in the Virginia Medical Monthly, June, 1877, where death followed the taking of 3 grams, divided into several doses, within a period of forty hours after the first dose. All these cases, however, are at the best doubtful, for in most instances the patient has had enough the matter with him to have killed him anyhow.

surreptitiously. In any case there can no possible harm result from labeling.

The same arguments may be repeated almost word for word in the case of copper.

Lead, tin, and zinc are not usually added intentionally, but are often present, and can not be otherwise described than as dangerous to health. Zinc is sometimes used as a substitute for copper in greening peas, but t comes into canned goods accidentally as a rule. Lead comes from the lavish use of solder rich in lead and from the use of low grades of tin plate. As to its dangerous nature there can be no question. Tin in many instances is almost unavoidably a constituent of canned goods where the common unvarnished cans are used. There are few samples of these goods in which it can not be detected.

SCOPE OF THE INVESTIGATION.

The directions given by the Chief Chemist for carrying out the work on canned vegetables provided that analyses be made of the commonly occurring brands in order to establish their nutritive value and that preservatives, metallic contaminations, and other foreign substances, be searched for. Furthermore, directions were given to examine a few of the tin cans to ascertain the quality of tin plate and solder in common use. In accordance with these instructions crude fiber, albuminoids, digestible albuminoids, ash, salt, fat, and carbohydrates were determined in each sample. The preservatives looked for were boric acid, salicylic acid, benzoic acid, sulphurous acid, saccharin, and hydronaphthol. In working upon metallic contaminations, copper, lead, tin, and zinc were searched for, and determined in many instances.

METHODS FOR PROXIMATE ANALYSIS.

GENERAL EXAMINATION.

The full can was weighed, opened, the juice poured off, and the can reweighed. The can was then completely emptied and once more weighed. The difference between the first and last weighings gave the total contents of the can; that between the second and third the solid contents, together with what moisture adhered thereto.

The moist solid matter was put into a mortar, thoroughly disintegrated, and mixed with the fluid. Portions of this mixture were dried to constant weight at 100° in a steam-heated oven. In the bath used this required about four hours. The loss gave the total water and other substances volatile at the temperature mentioned.

The rest of the pulped sample was placed on ordinary china plates and dried at 100° in a steam-heated bath. When the sample became dry enough to grind it was scraped off the plate, ground in a drug mill, and sifted through a sieve having holes 1 mm in diameter. All globules of solder were picked out as far as possible before grinding. Those which then escaped notice were flattened by the drug mill and easily picked out of the sieve.

The ground sample was bottled and used for analysis.

WATER.

In this powdered sample water was determined by drying at 100° in a platinum dish.

ETHER EXTRACT.

For the ether extract the sample used for water determination was extracted with anhydrous ether in a Knorr extractor, the ether driven off by a gentle heat, and the flask and extract dried to constant weight in a steam-jacketed bath. This required about three hours.

CRUDE FIBER.

The ether-extracted material was transferred to a flask and treated according to the method of the Association of Official Agricultural Chemists[1] for fiber. The solutions of acid and alkali were each 2.50 per cent.

ALBUMINOIDS.

Nitrogen was determined by the Kjeldahl method.[2] Albuminoids were calculated by multiplying the per cent of nitrogen by the customary factor, 6.25.

DIGESTIBLE ALBUMINOIDS.

Indigestible nitrogen was determined by Niebling's method,[3] viz: One gram of sample was washed with ether and the washed residue introduced into a flask, together with 50 cc of 0.2 per cent hydrochloric acid. The mixture was brought to a boil, allowed to cool, exactly neutralized, and 50 cc of Stutzer's pancreas solution added. The mixture was then kept at a temperature between 37° and 40° for six hours, shaking occasionally. At the end of this time it was cooled, allowed to settle for a few minutes, poured through a 14 cm Munktell filter, and the residue thoroughly washed with warm water (37° to 40°). The filter and residue were dried at 100°, introduced into a flask, and treated by the regular Kjeldahl process. The nitrogen thus found, multiplied by 6.25, gave "indigestible albuminoids," and subtracting this number from the total albuminoids found in the sample, left "digestible albuminoids."

The pancreas solution was made up according to Stutzer's directions,[4] as follows:

Free a cow's pancreas from fat as far as possible, mince, rub up with sand, and allow to stand in the air from twenty-four to thirty-six hours. Mix the mass with lime

---

[1] Bull. **35**, 215.
[2] Bull. **35**, 200.
[3] Landw. Jahrb. **19**, 149; abs. Chem. Centrbl., 1890, **2**, 116.
[4] Verssuchsst. **36**, 321; abs. Chem. Centrbl., 1890, **1**, 176.

water and glycerol, using for every kilo of fat-free pancreas 3 liters of lime water and 1 liter of glycerol, sp. g. 1.23. Add further a little chloroform, and set the mixture aside for from four to six days. Press the mass through a bag and filter the fluid through a coarse filter. Warm the fluid to 40° for two hours and refilter. The solution used in digestion is prepared by taking 250 cc of this extract and mixing with it 750 cc of water containing in solution 5 grams of sodium carbonate. Warm to 40° for two hours and filter. It is then ready for use.

ASH.

From the fact that canned goods always contain heavy metals, it was impracticable to determine the ash in the platinum ash dishes usually employed in this laboratory. An efficient and satisfactory substitute was found in the lids which are furnished with porcelain crucibles. Burning was done in a gas-heated muffle, on the floor of which was laid a piece of asbestos perforated at intervals. The rings were knocked off the lids and the slight residual projection was accommodated in the holes of the asbestos. Some little care is necessary in picking up the dish thus formed with the tongs. It is best to put one jaw of the tongs under the dish and the other above, thus picking it up sidewise, as from the slanting conformation of the sides of the dish an attempt to grasp it as an ordinary dish would be grasped is hazardous. These lid-dishes change very little in weight from use, and from their shallowness, permitting free access of air, a good ash is readily obtained. They are best marked with hydrofluoric acid. They seldom break.

SALT.

The ash obtained as above described was washed into a 100 cc flask with water and enough dilute nitric acid added to make the mixture blue congo paper. To this solution powdered calcium carbonate was added till there was a distinctly visible excess, and the solution boiled long enough to expel carbonic acid. After cooling it was made up to the 100 cc mark, filtered, and an aliquot part, usually 50 cc, measured into a beaker and titrated with tenth-normal silver solution, using potassium chromate as an indicator. Duplicates by this method are concordant.

"CORRECTED ASH."

In the tables the result found by subtracting the per cent of salt from the per cent of ash is recorded as "corrected ash."

PRESERVATIVES.

The use of preservatives is becoming quite common in the canneries. Some goods, corn for instance, are rather difficult to sterilize by short periods of heating, and with others heat exercises an influence upon the flavor or consistency, so that the addition of an antiseptic materially facilitates the canner's work.

If a can of food is heated to a temperature sufficient to kill all growing bacteria, the presence of an extremely small amount of a germicide like salicylic acid suffices to restrain all further fermentation, although the amount of antiseptic added might not have been sufficient to materially affect bacterial life if added to a solution in an active state of decomposition. Most of the bacteria commonly found will not resist a temperature of 65° to 70° when in the active state in a fermenting liquid, but these bacteria in the condition in which they are found in dust, or when in the shape of spores, resist this heat pretty well. If, however, the liquid in which these desiccated bacteria or spores occur contains a minimal amount of salicylic acid or other antiseptic, development into the vegetating form does not occur. Now, in exposing a can of food to the action of heat, no matter how conveyed, it is always a matter of difficulty to insure that the central portion of the contents of the can shall receive as much heat as the portions lying next the surface, and this is particularly true of solid-packed goods, such as corn and baked beans. It can be done in time, of course, but time is expensive. Dosing a food with a cheap antiseptic saves time and trouble and enables the canner to be quite certain of the keeping qualities of his goods, no matter in how slovenly or sloppy a manner his work may have been conducted. For this reason antiseptics are daily growing in favor among the preservers.

One objection to the use of chemical preservatives arises from the fact that they do not confine their anti-fermentative action to the food in the can, but continue to exercise it after the food reaches the stomach, which is not desirable. Digestion is effected by the action of unorganized ferments to a large extent, and on this action most antiseptics have a greater or less restraint.

It is difficult to say how far the use of preservatives cheapens canned goods. Of course all saving of labor or time tends to lessen the cost of production, but there seems to be no material difference in point of cost to the consumer between those brands of canned goods which contain antiseptics and those which do not. Probably were the use of preservatives discontinued there would be no material change in the retail price.

In the work done on the canned vegetables but two preservatives were found, if salt be disregarded, viz, salicylic and sulphurous acids. Salt is supposed to be added primarily as a condiment, and only secondarily as an antiseptic. It was present, however, in some cases in inordinate quantities. In one case (No. 10923) it constituted 40 per cent of the dry matter. Salicylic acid was found in 47.per cent of the total samples examined. Sulphurous acid was also very common.

METHOD FOR DETECTION OF PRESERVATIVES.

In the method[1] adopted for preservatives, the contents of the can

---

[1] This scheme is based in part on one proposed by the late I. T. Davis in an unpublished paper on meat preservatives.

were thoroughly pulped, about 50 grams mixed with dilute phosphoric acid, and the mixture allowed to stand for some time. The mixture was then strained through a coarse cotton bag. The resulting liquid, of which there should be about 50 to 75 cc, was then subjected to distillation. The first 5 cc of the distillate were examined for sulphurous acid. For this purpose add bromin water, boil till the yellow color disappears, and add solution of barium chlorid. Any precipitate of barium sulphate is of course indicative of sulphurous acid. A portion of the first distillates can also be tested for hydronaphthol by Beebe's method.[1] This is done by making the liquid very faintly alkaline with dilute ammonia, and then as slightly acid with dilute nitric acid. Next add a drop of concentrated solution of sodium nitrate. In the presence of hydronaphthol, a rose color is developed. The reaction is a delicate one, but the process requires much practice with known solutions before it can be used. The test, however, is extremely characteristic.

After testing for sulphurous acid and hydronaphthol, the distillation is allowed to proceed till the contents of the distilling flask are nearly down to dryness and salicylic acid tested for in the last portions, preferably in the last 10 cc. In case much is present salicylic acid comes over during the whole course of the distillation, but it is mainly contained in the final portions. This tendency may be illustrated by the following experiment: Twenty-five mg of salicylic acid were dissolved in 250 cc of water containing a little phosphoric acid and the mixture distilled, the distillate being collected in portions of 25 cc each. The first fraction gave a distinct but pale color with iron chlorid, the next a somewhat stronger reaction, and the next a still more marked color. In this last portion the salicylic acid amounted to about 0.3 mg. The next two portions gave increasingly brighter color tests, the amounts of the acid contained in each being, respectively, 0.4 and 0.5 mg. In the sixth portion there were about 0.8 mg, and in the following fraction about 0.9 mg. The eighth portion of 25 cc contained 2.2 mg, and a final portion of 15 cc contained 3 mg. The amount of salicylic acid in these distillates was estimated colorimetrically.

In detecting salicylic acid the method of macerating the contents of the can with dilute acid and distilling the resulting fluid directly works very well in those cases where the quantity of the acid present is not too small, but much better reactions can be obtained by drying the sample of canned food, powdering, making into stiff paste with sulphuric acid, extracting with ether, evaporating the ether, taking up the residue with water containing a little alkali, making acid with phosphoric acid, and then distilling. In the distillate thus obtained, salicylic acid can be distinguished even when existing in the original substance in quantities too small to be identified by the first mentioned course of procedure.

---

[1] Analyst, 1888, 13, 52.

The test finally used to identify the salicylic acid is preferably that with ferric chlorid, though the methyl ester reaction can also be used. The solution of ferric chlorid employed should contain about 5 mg to the cc and about 2 or 3 drops should be employed for each test.

In the ethereal extract obtained in the method for salicylic acid just given, saccharin if present may be identified by evaporating off the ether from a portion and tasting the residue. Should saccharin be present in the original sample in quantities sufficient to communicate to it a sweet taste, and it would hardly be worth while to add less, an unmistakable sweet taste will be obtained. There is no better method than this for saccharin, though many have been proposed. It is practically that used by the French customs officers.

Benzoic acid may be tested for by taking the unused portion of the distillate from the canned food sample, making alkaline with caustic soda, after adding a little silver sulphate, transferring to a porcelain dish and evaporating to dryness. Next heat on a sand bath to a fairly high temperature for about fifteen minutes. While still hot a little sulphuric acid is added, and the benzoic acid which is liberated, if present, recognized by the smell, or rather by its irritating effect on the nose. Considerable practice is necessary to use this test, but with pure chemicals as little as half a milligram can be recognized. Mohler's test[1] may also be used on the dried residue obtained after the evaporation. The reaction given by this test, however, is not characteristic for benzoic acid being also given by salicylic acid.

Besides the method already given for sulphurous acid, another was used frequently which is very convenient. In this method a portion of the liquid obtained by mixing the contents of the can with acid and straining off is mixed with hydrochloric acid, placed in a test tube, and powdered zinc added. It is then covered with a layer of ether and a piece of lead paper (paper moistened with lead acetate) laid above the mouth of the tube. In the presence of sulphurous acid, hydrogen sulphid is evolved and its presence shown by the paper assuming a brown color. The object of the ether is to keep down any frothing. Results obtained by this method agree very well with those furnished by the distillation method.

SULPHUROUS ACID.

This preservative, in the form of the fumes from burning sulphur, has been used from time immemorial as a general disinfectant and antiseptic. It exercises a bleaching as well as an anti-putrefactive action, and it is therefore greatly favored by corn canners.

Sulphurous acid, although not a normal constituent of food, is probably not directly harmful in itself. Its use, however, for foods put up in tin cans is to be deprecated for the reason that it attacks the tin and brings it into solution.

[1] See page 1167.

SALICYLIC ACID.

Salicylic acid was discovered in 1838 by Piria.[1] He prepared it by oxidizing the oil of *Spiræa ulmaria*. In 1843 Proctor[2] discovered it in oil of wintergreen, and Cahours[3] prepared it from this source in 1844. In 1852 it was synthetically made by Gerland.[4] In 1860 Kolbe and Lautemann[5] discovered a process for preparing it from carbolic acid, and in 1874 Kolbe[6] so improved the method as to render the acid commercially available. It is from this time that the use of the acid as a food preservative may be dated. Shortly after discovering his improved method for its preparation, Kolbe made an extensive study of the antifermentative action of salicylic acid which extended over the space of a year or two. He came to the conclusion that the acid restrained or prevented the action of organized ferments, and likewise that of unorganized ferments, to some extent, but that it was harmless to animal life. In the course of one series of experiments he took a daily dose of salicylic acid for over a year, commencing with half a gram and gradually increasing it to $1\frac{1}{2}$ grams daily. He reports his health to have been the same as usual during this experiment. He also administered the acid to others, and reports the same result. He strongly advocated its use as a food preservative.

Since that time the use of salicylic acid for this purpose has steadily increased, and there are probably now few canners who do not at least occasionally use it. The aggregate of the amount used yearly by the canners and sold for home use in the form of fruit preservatives must be very large. Most of the secret preservatives sold by the druggists and others owe their activity to its presence.

The use of salicylic acid as a food preservative has been forbidden by several European governments. France prohibited it in 1881, and renewed the prohibition in 1883.

An exhaustive discussion of the propriety of the use of salicylic acid as a food preservative took place at the Nuremberg meeting of the Freie Vereinigung der bayerischen Vertreter der angewandten Chemie, August 7 and 8, 1885. The association refused, by a practically unanimous vote, to sanction the addition of salicylic acid to beer. A special committee of the Paris Academy of Medicine[7] reported on this subject, that, while persons in good health might suffer no injury from the ingestion of such small amounts of salicylic acid as are liable to be contained in food, this did not necessarily hold good for the aged or for those in feeble health. Persons suffering from dyspepsia or diseased kidneys it was found were especially sensitive to the action of this substance. The report closed with a recommendation that the addition to food of

---

[1] Amer. J. Pharm., August, 1843.
[2] Ann. de chim. et de phys., 1838, **69**, 298.
[3] J. prakt. Chem., **29**, 197.
[4] Quarterly J. Chem. Soc., **5**, 133.
[5] Lieb. Ann., **115**, 201.
[6] J. prakt. Chem., 2, **10**, 93.
[7] Bull de l'Acad. de méd. (Paris), 1886, **16**, 582.

salicylic acid or its salts, even in small amounts, be absolutely prohibited.

Regarding the physiological effects of salicylic acid, the testimony is conflicting. There is a dearth of reliable experiments upon the human subject. As already mentioned, however, Kolbe took daily doses for the period of a year without injurious effect. Lehmann[1] administered to each of two Munich laborers half a gram of salicylic acid daily for seventy-five and ninety-one days, respectively, without a trace of injurious effect. These amounts are much larger than would ever be found in food. Administration of doses of salicylic acid, ranging between 6 and 12 grams, soon causes symptoms of cerebral poisoning. Four grams of sodium salicylate have been known to cause exceptionally severe toxic symptoms. The minimum dose for salicylic acid as given by the dispensatory is, on the authority of Ewald, 5 grams, repeated in five hours when necessary in cases where its antipyretic action is sought. Salicylic acid is one of the best known remedies for rheumatism in all cases where it is not directly contra indicated by renal affections. As to its influence on digestion, information is lacking. It is certainly not beneficial, however.

Its detection in food is fairly easy. It gives two very characteristic reactions. With ferric chlorid in nearly neutral solution it gives a deep-purple color, and treated so as to produce its methyl ester, a highly characteristic odor of wintergreen. It can be separated from food in a fairly pure state by acidulating the sample, extracting with ether, and distilling the extract in a current of steam.

Regarding the propriety of the use of salicylic acid by the canners, it may be said, as before remarked, that this use should be unhesitatingly condemned in cases where the fact is not indicated on the label of the goods. Salicylic acid may be harmless in very small doses to 99 out of 100 consumers, but the interests of the hundredth man should be guarded. Moreover, there is no safeguard against the use of inordinate quantities, for while the qualitative detection of salicylic acid is very easy, the quantitative estimation is a matter of very considerable difficulty. For this reason the canner who uses any at all may use almost any quantity he pleases with perfect impunity. Moderately large doses of salicylic acid are quite likely to prove detrimental to many people.

The whole salicylic acid question was quite thoroughly gone into in a previous part of this bulletin.[2]

SACCHARIN.

Saccharin was not found in any sample of canned goods. It is an article of too recent introduction to have found its way into many canneries. A sample of clear liquid in a bottle labeled "Superior sweetener,

---

[1] Methoden der praktischen Hygiene, Wiesbaden, 1890, 281.
[2] Bulletin 13, part 3, page 298.

Alex. Fries & Bros., 92 Reade St., New York," was sent in by Mr. H. E. Taylor, of 152 Clifton Place, Brooklyn, N. Y. The accompanying letter stated that this sweetener was used largely in the canneries of this country as an addition to canned corn, and was claimed by Fries & Bros. to be a good antiseptic, but perfectly harmless. On examination it proved to be a 12.8 per cent solution of saccharin.

As to the physiological action of saccharin little definite is known, although there is already a large literature pertaining to the subject. It seems probable, however, that in most cases it is not particularly deleterious to the human system.

BENZOIC ACID.

Benzoic acid was not found in any instance. The methods for its detection, however, are far too imperfect to allow the conclusion to be drawn that it is never used to preserve vegetable foods.

HYDRONAPHTHOL.

"Hydronaphthol" or beta-naphthol was not found in any sample.

BORIC ACID.

Neither boric acid nor borax was found in any sample. Both flame test and turmeric test were used. It is not probable that either antiseptic is used for canned vegetables.

TIN PLATE IN CANS.

Commercially tin plate is divided into two classes, known respectively as "bright" and "terne" plate, the former being covered with more or less pure tin and the latter with a varying mixture of lead and tin. "Bright" plate is, or should be, the kind used in canning food, terne-plate being intended for roofing. Inasmuch, however, as the price of tin is between four and five times that of lead, plate containing a little lead is somewhat cheaper, and there is a constant tendency among canners to use these cheap grades. In tinning iron plate with a mixture of lead and tin, the coating may be regarded as composed of two layers, the undermost being an alloy of iron and tin containing little lead and a surface layer of tin richer in lead than was the original metal used for plating. This concentration of lead on the surface is due to the fact that iron and lead have little affinity for one another.

Metallic tin placed in a weakly acid solution containing a small amount of lead throws down the latter in the metallic form, becoming itself oxidized. Similarly, if an alloy of lead and tin containing a small amount of lead is exposed to such an acid solution the tin dissolves first.[1] This play of affinities, however, does not hold good in the case of alloys containing large amounts of lead, such as solders,

---

O. Hehner, Analyst, 1880, 5, 218; G. Wolffhügel, Chem. Centrbl., 1887, 592.

for from these lead and tin may be simultaneously dissolved. In any case the difference in the affinities is too slight to allow the consumer to implicitly rely on it to save him from the effects of the canner's greed. Lead poisoning is a serious thing and no chances should be taken.

In Germany the laws relative to the composition of alloys used for plating tinware are rigid. The law of June 25, 1887,[1] reads:

"§ 1. Cooking, eating, and drinking vessels, as well as measuring vessels for fluids * * * shall not be tinned with an alloy containing more than 1 part of lead in 100 parts, nor be soldered with an alloy containing more than 10 parts of lead in 100 parts by weight.

"The tinning on the interior of cans for preserving food must satisfy the requirements of § 1."

The plating alloy of a few cans from the pea samples was examined. In most cases lead was present. The quantity found varied within tolerably wide limits, as may be seen from the subjoined table:

Per cent of lead in plating alloy.

Serial No.	Lead.	Serial No.	Lead.	Serial No.	Lead.	Serial No.	Lead.
10694	2.21	10706	6.63	10724	1.03	10889	0.00
10695	1.21	10708	13.03	10870	.53	10890	4.16
10696	1.08	10709	9.87	10871	0.00	10891	11.69
10697	0.00	10710	4.86	10872	.47	10898	0.00
10698	3.12	10711	.86	10873	.44	10900	5.30
10699	11.53	10715	.22	10874	0.00	10901	0.40
10700	12.42	10716	1.51	10876	.66	10905	0.88
10701	0.54	10717	1.41	10881	0.00	10906	1.64
10702	9.02	10719	.91	10882	3.26	10980	3.20
10703	0.40	10720	1.86	10884	1.68	10981	0.00
10704	0.00	10721	.41	10886	0.15	10984	0.62
10705	1.06	10722	.66	10883	0.00		

The results shown under Nos. 10699, 10702, and 10703 were all from cans packed by a single firm whose goods are widely sold in Washington. No. 10699 cost 13 cents per can and No. 10702 10 cents, but No. 10703 was sold at 18 cents, thus enabling the packer to use a better quality of tin. Nos. 10700, 10706, 10708, 10709, 10710, 10882, 10891, 10900, and 10980 are all bad. No. 10891 was put up in France.

In regard to the methods of analysis usually given for the detection of lead in tin plate and for the analysis of the coating alloy, it may be said that they are unsatisfactory. Pinette[2] proposes to attack the tin coating with dilute nitric acid, decant the fluid and suspended matter from the residual iron, evaporate to dryness on the water bath, take up with nitric acid, filter and weigh the tin oxid, and in the filtrate estimate the lead as sulphate. Results are calculated by adding the

---
[1] Quoted by J. Pinette, Chem. Ztg., 1891, 15, 1109.
[2] Chem. Ztg., 1891, 15, 1028.

lead and tin found together and ascertaining the per cent of the sum due to lead. This method suffers from the radical defect that alloys of tin and iron, such as are found under the outermost layer and in immediate contact with the iron, can not be parted by the usual nitric acid separation.[1] Tin goes into solution and iron remains with the tin oxid, from which it can not be extracted by nitric acid. In the tin oxid obtained by following Pinette's method the amount of ferric oxid ranged from 4 to 12 per cent of the total, running usually between 4 and 10. In filtering the tin oxid the nitric acid solution was difficult to filter clear and when wash water was used the filtrate became turbid. Of course all tin going into the filtrate would count as lead when using the customary sulphuric acid precipitation, and of course the error due to ferric oxid retained by the stannic oxid tends to counterbalance that due to dissolved tin, but for the increase in lead there is no such compensation. It may therefore be safely concluded that the method is not accurate.

In the analyses given the method used was to dissolve the plating with weak aqua regia, neutralize the solution with ammonia, add excess of ammonium sulphid, digest on the water bath, filter, once more digest the solid residue with sulphid, and refilter. In the united filtrates tin sulphid was precipitated by hydrochloric acid, filtered and weighed as stannic oxid. The solid residue from the sulphid separation was dissolved in nitric acid and lead determined as sulphate. This method, although accurate, was too cumbrous and time-consuming to permit the examination of many samples.

The tin-iron alloy is more difficult of solution than the surface alloy of tin and lead, so that in stripping the tin plate by acids, as in the methods just described, there is a tendency to leave this interior layer to a greater or less extent. For this reason, unless care in this respect be exercised, the subsequent analysis will show the ratio of lead to tin greater than it was in the original alloy used for tinning.

There is a qualitative method often proposed,[2] which consists in putting a drop of weak nitric acid on the tin, evaporating to dryness, and moistening the spot with solution of potassium iodid. A yellow coloration resulting is supposed to indicate lead and by its intensity to give an approximate idea of the amount present. An alloy of 99.5 parts tin and 0.5 parts lead thus treated gave a yellow color which did not materially differ from that given by an alloy of 88 parts tin and 12 parts lead.

USE OF SOLDER IN CANNING.

The analysis of the solder presented no difficulty. Weighed portions were treated with nitric acid and the tin oxid filtered and weighed. In the filtrate the lead was estimated as sulphate. Traces of copper, coming

---

[1] Fremy, Traite de chim., **3**, (pt. 1), 795.
[2] Perron, Chem. Centrbl., 1890, **1**, 731.

possibly from the soldering tools, were nearly always present, but were not estimated. All the solders examined were taken from the interior of the can, and were all from the pea samples.

Lead in solder from inside of can.

Serial No.	Lead, per cent.	Serial No.	Lead, per cent.	Serial No.	Lead, per cent.
10695	61.84	10711	51.86	10915	62.47
10699	57.64	10717	53.96	10916	63.73
10700	54.78	10719	55.98	10917	52.18
10702	58.58	10720	62.37	10918	51.03
10704	59.94	10910	60.31	10919	56.05
10706	60.34	10912	65.47	10920	60.64
10708	63.31	10913	53.87	10921	53.05
10709	43.60	10914	54.48	10990	53.07

It will be noticed that none of these samples approaches the German limit of 10 per cent. They are evidently "half and half" solder. It is said by the trade, however, that a 10 per cent solder is extremely difficult to use, owing to its infusibility. The German canners use it in compliance with the law, but dislike it and are resorting to many devices to avoid exposed solder. One way of doing this is to varnish the inside of the can and put on the top with a rubber joint.[1]

LEAD-TOPPED BOTTLES.

Several of the French samples were packed in glass bottles closed by a lead top. In view of the fact that the only assignable reason for the preference of bottles over the ordinary tin can is to avoid all danger of metallic contamination of food, this practice is a most extraordinary one. Sample No. 10885 may serve as an example of this method of packing. This bottle bore the inscription "Petit pois, extra fins, Dandicolle & Gaudin, Limited, Bordeaux, France," and cost 40 cents. The bottle itself was an ordinary white glass bottle with the top ground off. The cover was formed of a piece of sheet lead, fastened around the neck by an iron band. There was nothing whatever in the way of protection between the lead and the peas. Probably the packers went on the assumption that the bottle was not likely to get wrongside up during its travels from France to this country and thought the precaution superfluous. On analysis the metal was found to consist of 93.57 per cent lead and 6.43 per cent tin. Strangely enough the contents of the bottle were found to be almost free from lead. Copper there was in plenty, but little lead. Samples Nos. 10738, "haricots verts;" 10978, macédoine; 10979, Brussels sprouts; and 11146, asparagus, were all bottled by the same firm in a similar manner. Samples Nos. 10879, peas, and 10936, "haricots verts," were packed by Eugène Du Raix, also of Bordeaux, and were put up in a similar fashion.

---

[1] See page 1163.

METALLIC CONTAMINATIONS.

In searching for the metallic combinations of the samples of canned vegetables examined in this laboratory, the method generally used was as follows: As large a quantity as possible (50 grams when size of sample permitted) of the dried sample was burned or charred, this depending on the nature of the sample, in a capacious porcelain crucible. The charred mass was extracted with weak nitric acid in the cold, to avoid solution of organic matter, filtered, and washed. The mass of char and insoluble matter on the filter was dried, together with the filter, and then burned to a white ash in a porcelain crucible. This ash was transferred to a platinum dish and covered with a mixture of hydrofluoric acid and normal potassium fluorid.

It was then heated over a low flame till all the water was driven off, and then the heat pushed till the mass in the dish after fusing finally became infusible. The heat was then raised to redness, and kept at that point for a few minutes. After cooling, the fritted mass was treated with dilute sulphuric acid and heated till white fumes came off. It was next taken up with dilute hydrochloric acid and the solution added to that first obtained. From the mixed solutions copper, tin, and lead were thrown down, after adopting the usual precautions, by means of hydrogen sulphid. In the filtrate, zinc when present was determined by the Low ferrocyanid method (see page 1166). The mixed sulphids of lead, tin, and copper were treated with hot dilute solution of sodium sulphid, the resulting solution filtered, and the residue once more extracted in the same way. From the mixed filtrates tin was precipitated by hydrochloric acid, regulating the addition by congo paper. The tin sulphid was converted to oxid by careful ignition, and weighed in that shape. The residual sulphids of lead and copper were oxidized by nitric acid and the lead separated as sulphate or chromate. In the filtrate copper was determined electrolytically.

In many cases copper was determined colorimetrically, but lead was invariably estimated gravimetrically.

Tin and lead estimations in the case of canned goods are always subject to a grave element of doubt, from the fact that it is an extremely difficult thing to remove all traces of solder. Of course lead and tin in this shape are not particularly dangerous to health, or, at all events, not nearly so much so as are these metals when occurring in foods in the dissolved or combined state. Extreme care was taken in the effort to avoid this source of error, but the attempt can not have been always successful.

LEAD.

As just stated, the estimation of lead in canned goods is a matter of extreme difficulty—that is, the estimation of the lead existing as salts in contradistinction to that existing in the metallic form. Careless work on the part of the canners is responsible for the fact that many cans contain fragments of solder of varying sizes. Above a certain limit these particles can be hand picked and below this point in the method of grinding the samples used in the analyses of canned vegetables, much of the solder was flattened sufficiently to be picked out during sifting. Metallic lead is probably also present in another form in these goods. The food dissolves out more or less of the metal from the solder, and this is again precipitated in a finely divided state by the tin of the walls of the cans. This of course can not be mechanically separated. It is probable, however, that this latter metal, by reason of its state of aggregation, from a hygienic standpoint is about as dangerous as if it were oxidized. This is of course not true in the case of lead existing as solder. Solder, or metallic lead in any shape, present in food is bad, but in point of danger it is not comparable with lead existing as salts.

For the reasons adduced above, the quantities of lead found in the various samples of canned foods must be understood to represent merely the sum of the metal present in a finely divided condition and that in an oxidized condition. Were it otherwise, and did these figures represent altogether dissolved lead, canned goods would be a source of great danger to health. Of course, in the case of samples put up in bottles, solder is not present and the lead found was probably all in an oxidized condition.

Lead is a dangerous metal, and the canners are very free with it. The solder used seldom contains less than 50 per cent and it is found on the inside of the can in liberal quantities, not considering that present as detached particles. Besides this, the tin plate often carries lead to large extents. Were it not for the reaction cited in speaking of tinning alloys, that metallic tin precipitates lead from its solutions when present in large excess, careless canning would be much more dangerous than it is. Unluckily, because tin, although not innocuous by any means, is not so dangerous as lead, this reaction is not absolute. After a certain quantity of tin is dissolved lead begins to go into solution to some extent. The relative quantities of the two metals in solution will depend partly upon the nature of the food and partly upon the relative quantities of metallic lead and tin exposed to the action of the food. Some lead is frequently found dissolved, though not often where the canning has been conducted with proper care and good materials have been used.

Little need be said in regard to the poisonous nature of lead compounds. This is so well known that its repetition here is almost unnecessary.

The most unpleasant characteristic of lead is its property of accumulating in the system and then suddenly manifesting a strong poisonous action. "Wrist drop," "printer's palsy," "painter's palsy," and "lead colic" are common names for maladies produced by its action. The

ZINC AND TIN IN CANNED VEGETABLES. 1041

literature of medicine is full of fatal cases, though it is believed there are none recorded from lead in canned goods. Blythe states that in the five years ending in 1880 there were 324 deaths from lead poisoning registered in England. This is equal to about 20 per cent of the total recorded number of fatal cases from poisons of all kinds. But one was accidental and none were criminal.

Lead is the most dangerous from a toxicological point of view of the common metals. Its use in any way or place where it is liable to come into contact with food is to be earnestly condemned. Nearly all European states render such employment illegal.

ZINC.

Zinc is probably not often purposely introduced into canned goods. Zinc salts have been proposed for greening peas, and are said to be in use in France to some extent. The process is said to be a secret. It is not likely, however, that it is often used for this purpose. Around canneries the use of galvanized iron is very common and vegetable juices coming into contact with it could readily dissolve more or less zinc, for that metal is quite soluble in acids, even when weak. Its introduction into canned goods might also happen from the use of zinc chlorid as an aid in the soldering operation.

Toxicologically zinc resembles copper in that, while an emetic in large doses, small quantities are not known to be specially poisonous. The dose of sulphate given as an emetic is about 1.3 grams (equivalent to 294 mg of zinc). For this purpose the salt is regarded as one of the best known. Many cases of fatal poisoning are on record from zinc salts, but these are all from the use of large quantities. Ordinarily, however, a large dose of a zinc compound is not dangerous, for the reason that the stomach at once rejects it. Little is known relative to the effect of small amounts of zinc, continually administered for long periods of time.

There is no legitimate reason for this metal being in canned goods and its presence there is usually the result of gross carelessness. It was found in many of the samples of canned vegetables examined and occasionally in relatively large quantity. In the pea sample, No. 10629, packed at Bordeaux by Vve. Garres & Fils, and bought in Florida, it was present to the extent of 85.5 mg per kilo. This may be one of the samples greened by the zinc method just referred to. The peas were bright green.

TIN.

Of the possible metallic contaminations, that caused by tin is, next to that of lead, probably the worst. It is assuredly the most common. In every sample of canned goods, which has been put up for any length of time, tin may be found dissolved or rather present in the oxidized or combined state. Varnishing the inside of the can, which is sometimes done, hinders this solution of the tin in great measure, but generally there is sufficient metal left bare to allow appreciable quantities to dissolve. The quantity which is taken up from a bare can is of course almost entirely dependent upon the nature of the contents. Highly acid liquids like those which surround some canned fruits dissolve the greatest quantity. Goods which have been sulphured also act heavily upon tin, forming tin sulphid, which probably then slowly dissolves to some extent. This appearance may be often noticed on the inside of cans which contain corn. The tin sulphid forms soft pasty deposits on the metal, looking very much like mold to the naked eye. Indeed, these spots are very often taken for fungoid growths. The tin sulphid in this form is readily rubbed off into the food by any shaking of the can.

The formation of the sulphid occurs by virtue of the fact that metals very readily reduce sulphur dioxid, forming a metallic sulphid or hydrogen sulphid, according to circumstances.

Tin poisoning resulting from the use of canned foods is not often recorded, although it is probable that minor disturbances to health frequently occur from this cause.

In 1880 O. Hehner examined a large number of canned foods, representing most of those thus preserved, for dissolved tin. In most of these it was found. Experiments made on Guinea pigs with doses of stannous hydrate showed a marked poisonous and in many cases a fatal action. This is the state in which tin occurs in foods. Stannic hydrate was found to be relatively harmless. The doses of tin which he employed, though not large in themselves, were, considering the size of a Guinea pig, rather heavy. Still, his conclusion that tin has well-marked toxic properties seems well justified. An abstract of this work is given on page 1164.

In 1883 Ungar and Bodländer examined a number of samples of canned goods, mainly asparagus. There had been a number of cases of illness resulting from the use of canned asparagus, which led them to make the investigation. Most of the foods examined contained the metal in greater or less quantity. In 1887 they resumed the subject and made a number of experiments upon the toxic properties of tin, using animals. An abstract of this work will be found on page 1165. Their conclusion was that tin in the stannous form, the one in which it occurs in canned foods, is capable of causing disturbance, more or less grave, when swallowed, and that chronic tin poisoning might possibly occur. They also proved that tin present in food in an insoluble form was dissolved and absorbed during digestion.

GREENING VEGETABLES WITH SALTS OF COPPER.

Copper in varying proportions is found in many samples of canned goods. It is derived to some extent from copper pans and other utensils of the canneries, and to some extent it is added directly in the form of soluble copper salts. Its natural occurrence, at least occasionally and in minute quantities, is also probable. Green vegetables in the presence of minimum amounts of copper compounds do not turn yellow in cooking, but preserve a fresh, green color. This phenomenon is well known to all canners and has been in use since the beginning of the art. For greening pickles, the use of copper kettles has been known for centuries. The cause of the phenomenon has never been satisfactorily elucidated, though Tschirch has proposed an explanation.[1] This is that in the process of cooking the chlorophyl is converted by the weak vegetable acids into two bodies, one of a basic nature and one of an acid—phyllocyanic acid. By this change the vegetables lose their original color and become brown. If, however, a salt of copper is present this combines with the phyllocyanic acid, forming a body of intense tinctorial power, being comparable in this respect with eosin. Solutions of one part in 200,000 show a blue color. The salt is soluble in alcohol, but not in water or dilute acids. It contains 9.2 per cent of copper oxid. Tschirch, from physiological investigations, was of the opinion that the compound is not harmful and that the presence in canned goods of 100 mg of copper oxid in the form of the alcohol-soluble salt should be made legally allowable.

Since the earliest development of the art of canning, it has been a maxim in the canneries that in order to secure the best results green peas should always be cooked in copper kettles and that the kettles should not be too clean. Vegetable juices, even when only faintly acid, exercise a remarkable solvent action on copper oxid, and, in turn, in their presence copper oxidizes readily; so that when cooking is done in copper kettles and too much cleanliness is not exercised, vegetables readily absorb enough copper to produce the desired effect. The amount necessary is very small, only 20 to 30 mg per kilo, or, in other words, 20 to 30 parts per million, of the vegetable. Lately, that is, within a few years, the preservers have learned that copper kettles are not necessary, but that the desired effect can be as well produced by the direct addition of copper sulphate or acetate, and this is now the usual practice. In 1890 an Alsatian firm (J. Clot & Cie) patented a method of greening peas by which the copper vessel containing them was placed in connection with one wire from a dynamo, while the other wire was connected with an electrode hanging in the fluid surrounding the food. The current passing through the mixture dissolved enough copper to color the peas. A translation of this patent is given on page 1162.

For many years there has been a dispute as to whether copper might be called a normal constituent of food. Many cases of its occurrence in the animal kingdom are known, and it is of course pretty certain, in view of this fact, that it must be a constituent of some vegetable tissues. One noteworthy occurrence of copper in the higher animals was discovered by Prof. A. H. Church in 1869.[2] The wing feathers of a

---

[1] Ztschr. Nahr. Hyg., 1891, 5, 221.
[2] Chem. News, Amer. Rep. 1869, 5, 61; Chem. News, 1892, 65, 218.

number of species (18 out of 25 known) of the turaco, or plantain eater, an African bird, are colored red by a pigment containing about 6 per cent of copper. These birds undoubtedly derive their copper from their food, and from its constant occurrence in their plumage it must be a normal constituent of the vegetable foods upon which they live.

In 1858 A. Dupré and Dr. Odling[1] investigated a large number of common vegetable materials and foods for copper. In nearly all cases the metal was found, though always in very minute quantity. Out of 22 samples of bread 21 contained traces; 20 samples of flour were all found to contain it, and the same was the case with 43 miscellaneous samples, comprising wheat, barley, maize, wheat straw, barley straw, turnips of different kinds, and beets. In 25 samples the metal was estimated. The maximum amount found was in a sample of wheat ash, being 0.024 part in 100 parts of ash. This corresponds to about 1,000 parts of fresh wheat, equivalent to about one part of copper oxid to 240,000 parts of wheat, in turn equivalent to something over 3 mg of copper per kilo. A sample of turnip contained the minimum amount determined, the copper as copper oxid amounting to about 1 part to 4,375,000 of the fresh turnip. When enough material was taken the copper was almost always found. That naturally present, however, seldom ever amounted to more than 1 part to 200,000 (5 mg per kilo). Animal substances (29 samples) were also tested and nearly all yielded the metal. Human liver showed 2 parts of copper oxid per million; sheeps' liver, in two cases, 50 parts; and kidneys, about 10 parts. Dupré and Odling came to the conclusion that in the case of vegetable foods the presence of more than 10 parts of copper per million of food must be looked upon as an adulteration. For green peas, however, they placed the limit at 18 to 36 parts. The method of analysis employed was to precipitate the copper on a platinum wire by a weak galvanic current, redissolve in nitric acid, and finally weigh as copper oxid.

Many other chemists have sought to show the constant existence of copper in animal and vegetable tissues, and a fairly large literature on the subject has accumulated. It seems to be proved that copper often occurs, but the proof adduced does not show at all that the metal is a constant component of the tissues in question. The mere fact that it is not found in many cases proves that it is not a normal constituent of living tissues. The drift of the evidence seems to be that it was an accidental ingredient in many cases where it has been found in tissues from the higher animals and plants. That it is not always so is shown by the case of the turaco.

Copper is present in many soils, and since the metal forms no salts which are particularly insoluble, it must often happen that plant roots find dissolved copper presented to them, and no doubt they sometimes absorb it. The amounts of copper found in common plants, however,

---

Guy's Hospital Reports, 1858, 103, Analyst, 1878, 2, 1.

so far as known, for a systematic search for the metal throughout the vegetable kingdom has never been made, do not amount to more than the veriest traces. In the work done on the question of the natural occurrence of copper it is a noticeable fact that fresh, unmanipulated vegetable products invariably have yielded the lowest amounts and that the highest have been found in such things as chocolate and other much-handled substances. In animals it is natural to expect copper to be more abundant than in plants, since copper undeniably has a tendency to accumulate in the system to some extent. One of its most usual depositing places is the liver.

All these facts, however, relative to the frequent occurrence of small amounts of copper in unexpected places have no bearing on the question whether the metal is injurious to health or not. If it is injurious to health, the fact that it sometimes occurs naturally in wheat is to be deplored, but is no excuse for artificially introducing it into another food. If it is not injurious, the addition would be less culpable, but by no means laudable. Copper is certainly not needed by the human system, and when introduced the organism can make no use of it, but sets to work at once to expel as much as possible of it.

As to whether the metal in small doses is injurious or not, this is still a debatable question. It seems quite evident that it is not nearly so poisonous as it was once reputed. Since large doses, however, unquestionably produce markedly unpleasant, though possibly not deadly, effects, it seems a plausible conclusion that small ones are not altogether innocuous.

Galippe made careful experiments, first on dogs and subsequently on himself and family, as to the effect of daily repeated small doses of copper, and came finally to the conclusion that they were not injurious. His experiments have been tried by others, though not on so extensive a scale, and as a general thing with the same result. He, however, has been trying to do a hard thing—prove a negative. The mere fact that one man can bear a certain medicine very well by no means proves that his neighbor is equally insusceptible. Common experience is sufficient to show this. One man habitually uses tobacco; to another it is an acrid poison. Granting for a moment that copper is injurious, it might very well follow that M. Galippe and his fellow experimenters formed a copper habit, and acquired the power of tolerating the metal in a manner analogous to the way in which toleration for many notorious poisons—arsenic or morphin, for instance—can be attained. It is not contended, of course, that this was necessarily so, but the fact remains that M. Galippe's results, which he interprets to mean that copper is absolutely non-poisonous, may be interpreted in other ways.

On the whole, it may be said that the question of the poisonous nature of copper salts is still an open one, and science is not yet ready to form an opinion. This being the case, it is believed that the practice of coppering peas, even when it is made known to the consumer, should be discouraged, and surreptitious coppering should be repressed.

The consumer should have all the facts before him, and if he then elects to run the possible risks there is no good reason in the present state of knowledge for restraining him. It is not, however, fair or good policy to make him take through ignorance greater or less amounts of a possibly or probably deleterious substance.

COPPER-GREENING IN FRANCE.

The whole question has probably been more extensively investigated in France than elsewhere. As is well known, in that country the preserving industry is one of the best developed, and the quantity of canned and bottled foods yearly exported reaches enormous proportions. For this export trade the United States is one of the most important markets. Of the foods thus preserved canned leguminous vegetables form by far the greater proportion, and of these it is estimated by the French that 90 per cent are artificially greened, mostly by copper. This trade is comparatively old and is firmly established. Peas, beans, etc., when coppered assume a peculiar green color not altered by cooking, and resembling the color of the raw vegetable, though not at all like the natural color of the fresh-cooked legumes, and a taste for goods of this color has grown up.

The preserving of goods thus colored has, until within a few years, been prohibited in France. It is true the law has been a dead letter, but it has none the less been on the statute books. Of course, the prohibition has not been a pleasing one for the canners, and they have been persistent in their efforts for its abolition. This, however, they were not able to secure from the government for many years in spite of the influence which so great an industry must necessarily have had. Anxious to please them, the Government has appointed commission after commission of scientific men to investigate the copper question with a view to rescinding the prohibition, should these investigations show just ground for such action. Uniformly, up to 1889, however, these commissions reported adversely, and the Government could not take the desired action. In all, the struggle between the scientific Dr. Jekyll and the commercial Mr. Hyde in France lasted nearly half a century before culminating in the victory of the latter. The history of the contest is an interesting one.

In Paris, a police ordinance,[1] dated February 28, 1853, prohibited the use of vessels and salts of copper in the preparation of food.

In 1860 the Comité consultatif d'hygiène publique on the report of a committee composed of MM. Bussy, Ville, and Tardieu, recommended that this local law be made applicable to all France. The report in substance, read as follows:[2]

The fact of the introduction of copper salts into preserved green vegetables and fruits has been proved. Though the amounts present are not sufficient, generally

---

[1] Recueil des trav. du Comité consultatif d'hygiène publique, 1875, 358.
[2] Gautier. Conserves alimentaires reverdies au cuivre. Ann. d'hyg. publ., 1879, s. 3, 1, 14.

speaking, to produce serious accidents, yet the addition of an eminently poisonous substance to food, in indeterminate proportions, constitutes a danger which can not be ignored and which the Government should not tolerate. The committee does not hesitate to condemn the use of vessels and salts of copper for preserving fruits and leguminous foods.

This resulted in a ministerial decree, dated December 20, 1860, carrying this recommendation into effect February 1, 1861.[1] It read as follows:

> In consideration of the fact that the employment of vessels of copper, or the addition of copper salts, in preserving fruits and legumes, presents danger to the public health, and in consideration also of the fact that the prohibition of this practice, which is rendered necessary to protect the health of the consumers, will not be injurious to the welfare of the industry, inasmuch as other greening methods exist:
> 1. Manufacturers are forbidden to use vessels or salts of copper in the preservation of preserved fruits or legumes intended for food.
> 2. Violations will be prosecuted before the courts and punished according to law.

This law, however, remained a dead letter for many years, though the Parisian police went through the form of buying and examining samples in 1869, 1872, 1873, and 1875, finding no copper, however.

In 1876 public and official attention was once more called to the subject.

In that year Dr. Micé was commissioned by the prefect of the Gironde to investigate copper greening, a manufacturer having asked permission to put up copper-greened vegetables destined for export. The report[2] of Dr. Micé was unfavorable and permission was refused.

About the same time P. Carles of Bordeaux examined a number of samples of canned vegetables for a Bordeaux merchant, who was anxious to send his customers in foreign countries only irreproachable goods and in whom the bright green color of the goods he had been selling had awakened suspicion. The examination showed that they contained copper in considerable quantities. From one can were obtained 49 mg of copper oxide, equivalent to 155 mg of the sulphate. The conclusion of M. Carles, however, was:

> As a can is ordinarily divided up among several persons, this quantity would not appear to present serious danger.

In the same year[3] a manufacturer of canned foods called the attention of the Parisian police to the fact that foods greened by copper were on sale in Paris, and stated that this was dangerous to health. The police bought 18 samples of the goods in question and submitted them to Pasteur for analysis. This chemist reported the presence of copper in 10 samples.

In his report (February 8, 1877) to the Conseil d'hygiène publique et de salubrité, he stated as follows:

---

[1] Brouardel, Ann. d'hyg., 1880, s. 3, 3, 198.
[2] Trav. du Conseil de salubrité de la Gironde, 1876; quoted by A. Chevallier, Rep. de pharm. (Fr.), 1877, 5, 371.
[3] Ann. d'hyg. publ., 1880, s. 3, 3, 199.

The presence or absence of copper in preserved peas can be readily ascertained by simple inspection. If the peas present the slightest green tint copper is present; if they are of a yellowish hue unmixed with green copper is absent.

From my own observations, as well as from reliable testimony from another source, I can confidently state that, in the present state of the preserving art, it is impossible to put up peas of a green color without recourse to the salts of copper.

The liquid surrounding copper-greened peas contains very little copper and the bulk of the metal is contained in an insoluble state in the peas, lying for the most part in the outer layers.

In those samples examined which contained the most copper, the quantity did not reach one ten-thousandth (100 mg per kilo) of the weight of the peas, exclusive of the surrounding liquor.

Dr. Galippe, in his study of the toxicity of the salts of copper, came to the conclusion that they were not poisonous. Granting that later experience confirms his results, the government should none the less proscribe the treatment of food with copper salts. He who asks for peas (petits pois) asks for a natural product of the vegetable world, from which copper is absent. Toleration of copper-greening should not be granted, save on the condition that the manufacturer and retailer legibly mark the packages containing peas thus treated: "Peas greened by copper salts" ("conserves de petits pois verdis par les sels de cuivre"). In this case toleration amounts to prohibition, since it is not likely that the consumer would buy goods bearing this inscription.

The Conseil adopted his report at the meeting on February 9, 1877.

M. Pasteur's letter was submitted to the minister of agriculture and commerce by the Paris prefect of police, May 17, 1877, together with a statement from the president of the association of Parisian canners, a M. Dumagnou, saying in effect, first, that experience had demonstrated that copper in the quantities used was harmless and, secondly, that the rigorous enforcement of the law of February 1, 1861, by interdicting the practice of copper greening, would effect the complete ruin of the preserving industry.

On receipt of this communication the minister laid the question before the Comité consultatif d'hygiène, which appointed another commission, this time composed of MM. Bussy, Fauvel and Bergeron. M. Bussy had been a member of the 1860 committee. On July 15, 1877, Bussy submitted a report, reaffirming the conclusions of the former commissions. Abstracted the report reads as follows:[1]

The leading argument of the manufacturers who are using the copper process of greening peas is that the consumers prefer those brands in which copper is used, and that this fact has forced the use of the process. Doubtless it is true that buyers prefer "green peas" but not "peas greened with copper sulphate," as is shown by the statement of the manufacturers themselves, that not a can would be sold, if it were so labeled. It is an adulteration by which the buyer is forced to take, contrary to his intentions, a food which he would regard as prejudicial to health if informed of its true character. Furthermore, it is contended that the amount used is not dangerous, but who can answer for the innocuousness of a possibly poisonous substance, administered daily for indefinite periods? Who can give a safe limit, taking into account varying ages, constitutions, and states of health of possible consumers? Grant that a limit could be fixed, would the manufacturers who have violated the existing law—a matter easily detected—be more likely to observe a law whose vio-

[1] Recueil des trav. du Comité consultatif d'hygiène publique, 1877, 7, 302.

lation would be far more difficult of detection? Who will answer for the mistakes and carelessness which pertain to every industrial process?

From the point of view regarding public health, two considerations dominate all others: (1) salts of copper are poisonous; (2) their addition imparts no useful quality to the food, but only a factitious color intended to mislead the consumer. Even if there were no other methods of preparing these foods than this, this one would be inadmissible—but there are others.

If those now known do not prove satisfactory, and the demand for green vegetables should still continue, it is not too much to require of the science and industry of our packers that they should find a process for securing the desired results without the use of questionable means.

So far from the practice tending to promote foreign commerce, it is injuring it. In England dealers are already being prosecuted for selling French peas. The suspicion cast upon the products will, if perpetuated, certainly work only harm to our foreign trade.

Can the Government, which, in the interests of foreign commerce and of the health of children, pushes its foresight so far as to forbid the sale of toys colored with poisonous pigments, sanction the staining of a common article of food with sulphate of copper?

The report was accepted and the prohibition of the use of copper renewed.

M. Bussy again appears on the scene in the next year, 1878, as a member of another commission to investigate the copper question, his colleagues this time being MM. Wurtz and Gavarret.

It appears that MM. Lecourt, a canner, and Guillemare, a professor of chemistry at Rheims, addressed a joint petition to the minister of agriculture and commerce. The substance of the document was that the petitioners had discovered a method of greening peas not dependent upon the use of copper salts, and wished the full rigor of the law brought to bear upon the other canners who still used copper. Abbreviating this petition it reads as follows:[1]

We have found a method capable of imparting to leguminous preserves the green color required by commerce, yet not involving any dangerous substance. The process is applicable on the large scale, is salubrious, and does not, like the present process, involve the salts of copper so justly prohibited by law. We demand in consequence that the administration, after convincing itself of the value of the proposed process, in the interest not only of justice but of public hygiene, rigorously enforce the laws against the employment of copper salts for greening vegetables.

Ought we, in the presence of a tolerance which permits our competitors to manufacture vegetables greened by the inexpensive copper process—a practice which is, according to the Comité d'hygiène, a true fraud—ought we to persevere in the use of a process which, under these conditions, is unprofitable and irksome? We think not.

We beg the minister to submit our process of greening by chlorophyl to the Conseil d'hygiène. We believe that our method would meet with approbation. We would further ask that the laws which regulate the matter of copper greening be rigorously enforced.

The petitioners' process consisted in the addition to the vegetables to be greened of an alkaline solution of chlorophyl, preferably extracted from spinach.

---

[1] Recueil des trav. du Comité consultatif d'hygiène publique, 1878, 8, 366.

The petition was referred by the minister to the Comité which appointed the commission just named. A report was rendered December 30, 1878. The result of its deliberations may be summed up as follows:

The process submitted by MM. Lecourt and Guillemare gives satisfactory results, and from a hygienic standpoint is free from the objections which have been urged against the use of copper and its salts.

However, while awarding commendation to the petitioners, the commission regard it as desirable that the Comité should express the opinion that, no matter how inoffensive the means employed, it is desirable that the manufacturers renounce all artificial methods of coloring. It is believed that the public would prefer vegetables of good quality with their natural color to those presenting the seductive green which could be a means of concealing inferior quality.

This progress in the methods of food preserving, however, furnishes a further reason for a more vigorous enforcement of the laws prohibiting the use of salts of copper for coloring vegetables.

The commendation of the administration should be expressed to the petitioners for the service they have rendered public health.

This was not altogether unmixed commendation.

The next appearance of the question is due to MM. Bouchardat and Gautier, who were appointed a commission to investigate the general subject of the artificial coloring of articles of food and drink, and the dangers resulting therefrom, by the organizing committee of the International Congress of Hygiene, which met in Paris in 1878. That portion of their report as delivered to the congress which deals with the subject of artificial greening of preserved vegetables is as follows:[1]

The packing of canned vegetables may be called a French industry. Twenty to twenty-two million packages (demi-boîtes) of peas, green beans, flageolets, etc., are annually packed, 90 per cent of which are exported.

Copper-greening of these vegetables has been in use for twenty-five to thirty years and is practiced by nine-tenths of the packers. It originated in the observation that vegetables cooked in copper kettles preserved their color.

The copper absorbed is deposited in the outer layers of the vegetables, forming a blue albuminate. The color of this deposit, mixed with the yellow produced by the cookery in the vegetable, forms the green tint so much desired.

We believe that copper acts largely by virtue of its antiseptic and antifermentative properties, opposing the action of diastase which would tend to destroy the chlorophyl. To confirm this we may cite the fact that metals giving colorless albuminates also preserve the color. This is true with zinc and mercury.

Legumes preserved by the copper process always contain this metal, sometimes in notable quantities and sometimes in mere traces.

Nine-tenths of the preserved green vegetables sold in France or foreign countries are greened by copper. The process is in use at present in Germany, Italy and Spain. By virtue of a general understanding, every can of legumes sold in France and not bearing the inscription "Legumes au naturel" is known to be greened with copper.

There are, however, other processes in use.

Coloring by a lake of chlorophyl is one, the process in which lime sucrate is used is another, the Garges method a third, and, finally and lastly, zinc salts are sometimes used.

---

[1] Comptes rendus stenog. Congres Internat. d'hyg., tenu á Paris, 1-10 aôut, 1878; Paris, 1880, [10], 1, 501.

In the first-named process, chlorophyl is extracted from spinach or nettle with caustic soda, precipitated in the form of a lake by potash alum, this dissolved in sodium phosphate and the legumes plunged into the hot solution, from which they take up the coloring matter. This process was patented in 1876 by MM. Lecourt and Guillemare, who amended it in February, 1877, by proposing to replace the alum by soluble lime and magnesium salts. Coloring by this process is irregular. Some packages contain legumes which are of a good green color, but with others this is not the case. Furthermore, the vegetables thus prepared have lost their own delicate flavor to assume that of the nettle or, worse, that of the spinach, which served as a source of chlorophyl. Besides all this the method is long and delicate and difficult of execution. Again, the goods do not keep well, and the cans are liable to explode, from internal fermentation.

The calcium sucrate method was patented August 8, 1877, by MM. Possoz, Biardot, and Lecuyer. They add to the can, before the final cooking, a liquid composed of 100 parts water, 3 parts sugar, 1 part sea salt, and 0.4 parts lime, with the idea of precipitating a comparatively stable lime lake in the tissues of the vegetable. The process gives variable results and has yet to stand the test of time. The color of goods thus preserved is yellowish green.

Garges' patent was taken out September 4, 1877. He treats the vegetables first with sodium carbonate, then with alum. The results are unsatisfactory.

The zinc process is still kept a secret, but in principle it consists in substituting zinc chlorid for the copper sulphate used in the common process. It is in use by one of the large factories. Vegetables treated by it assume a green tint or a natural yellowish green, but they do not assume the French green, which the consumer unhappily associates with these goods. This method of greening in our opinion should be suppressed, since zinc is a substance whose presence in food in quantities as large as those in which copper is customarily employed, can not be tolerated without risk to public health.

The liquor which surrounds the canned green peas, beans, etc., ordinarily contains no copper. In some factories, however, it is a custom to add a little copper sulphate to it, which is a reprehensible practice. The copper in the legumes is not soluble in boiling water, but if artificially digested, copper passes into solution.

A part of the copper dissolved in digestion is absorbed and part passes into the excrement. It is necessary to consider the effect of the introduction into the system of small and repeated doses of a metal reputed to be dangerous. In France, the Conseil d'hygiene, consulted upon this point, has not hesitated to condemn the use of copper salts, and the administration has accordingly taken measures to prevent it.

The salts of copper are poisonous: they are violent emetics. It is not easy, however, to poison an animal capable of vomiting. Numerous experiments[1] show that a dog or even a man can swallow daily several grams of acetate, sulphate, phosphate, or iodid of copper without experiencing any greater discomfort than a temporary colic or fit of vomiting. However, it is not known whether small doses taken daily can cause even these slight symptoms.

Copper colic is described by some writers, admitted in rare cases by others, and by most authors totally denied. To-day it may be said that it is admitted to be an unimportant phenomenon. Workers in copper absorb the metal directly by contact and also in the state of dust. Often they become so saturated with it that their eyes are turned green and the urine contains it. Galippe and his family lived for more than a year on foods prepared in copper vessels and suffered no bad results.

It would seem that in doses in which the metallic taste is not perceptible and in which there is no emetic action the ingestion of copper salts leads to no immediate inconvenience; but it also seems that a longer experience and more rigorous observa-

---

[1] Toussaint, Bul. thérap., 55, 237. Galippe, Theses de Paris, 1875.

tion of statistics is necessary before the absolute innocuousness of oft-repeated small doses can be declared. It appears a logical conclusion that the introduction of copper in the greening process is to be viewed with suspicion, and therefore prohibited; but this conclusion would have less weight were it shown that copper existed in equal quantity in any of the usual foods of which experience has shown the innocuousness.

The quantity of copper found by different chemists in various samples of peas, beans, etc., varies widely, running all the way from 20 to 222 mg of copper per kilo, this being a tolerably wide range. In many instances more was found than was requisite for coloring purposes. The results would probably have reached a higher figure had "haricots verts," "ecossées," or "flageolets" been added to the list of those samples examined. The variations may be partly due to the different absorbent powers of the different vegetables. It is also due in part to the blameworthy practice of some canners of introducing a small quantity of copper sulphate into the liquor in the can just before sealing. Pasteur found a maximum of 100 mg of copper per kilo of vegetables (after pouring off the juice); Bussy found 10 mg; Galippe, 60; and Carles, 210 mg.

In the course of our own work on this subject we wished not merely to determine the copper, but also to answer the serious question whether lead is introduced into the animal economy by means of the tinning and solder habitually employed by the canners.

In determining the heavy metals by the method adopted the dried substance moistened with nitric acid is carbonized in a platinum dish at a low temperature over a glass lamp and in a room free from floating dust. The char is then finely ground and treated with water acidulated with nitric acid. The carbon filtered off is easily burnt at a low red heat. The filtrate and washings are evaporated to dryness. They do not generally contain any copper. To this residue is added the ash resulting from the burning of the carbon. The mixture is heated till no more nitric fumes are given off, burned, water added, boiled and cooled, and filtered after twenty-four hours. The lead and tin remain on the filter and the copper passes into the filtrate in the form of sulphate. This is easily precipitated with a two-cell Bunsen battery, the filtrate being rendered moderately acid. Wash at the end of forty-eight hours by successive decantations and weigh as metallic copper. Lead remains as sulphate, tin as metastannic acid, on the filter. Boil the residue several hours with a little crystallized barium hydrate. The lead passes into the state either of hydrate or of barium plumbate; tin becomes a stannate. Treat with hydrochloric acid, heat and filter through glass wool. Wash with boiling water, acidulated with hydrochloric acid, to extract all the lead chlorid, and mix the filtrate with boiling hydrochloric acid. From this solution tin and lead are precipitated by hydrogen sulphid. The sulphids are collected on a filter, washed with water charged with hydrogen sulphid, and digested in a little dilute, tepid solution of alkaline polysulphid, which dissolves out the tin and leaves the lead. Precipitate the tin from the resulting solution by a few drops of acid and calcine the sulphid after having oxidized it with nitric acid to some extent. The residual lead sulphid from the separation is transformed into sulphate by nitric acid and weighed as that salt.

The maximum amount of tin found was 71 mg; of lead 7.7 mg per kilo. It may be said of tin, as of copper, that its action on the animal economy is uncertain in such small doses, but in the case of lead, even the small portions found in the vegetables examined must be regarded as a serious matter. The source of lead is generally the solder. In making a tin can there are left three lines of solder, one at each end and another down the center of the can. Pieces of solder are also often dropped in in sealing the can, and this increases the danger. This solder consists of two parts lead and one part tin. A further aggravation of the dangers existing from these sources is found in the employment of cheap tin plate, the cheapness of which is in direct ratio to the amount of lead found in the tinning alloy.

Action on this matter is strongly urged, and the following addition to the ordinance of 1853 (that relative to the use of copper vessels) is suggested:

"The use of lead, of zinc and of galvanized iron is forbidden in the manufacture of articles destined to contain alimentary substances or for drinking vessels."

The claims urged by the canners in defense of the practice of coppering may be summed up: The practice has existed 28 years; it is used with 95 per cent of the legumes preserved and no accident has happened from their consumption in the history of the industry. The laborers employed in the canneries consume them in large quantity for several months in the year with entire impunity. The consumers prefer green vegetables to those *au naturel*. The demand both at home and abroad for those brands prepared with copper almost to the exclusion of the others, proves that they are not only safe but more pleasing, and it is this demand which has little by little universalized the practice of greening. If the manufacturer does not put copper in, the cook will and it is better that it should be added under the strict surveillance which an extended industry demands than that it should be left in incompetent hands. The addition of too much copper is impossible. The bad taste resulting from too large a dose furnishes sufficient guaranty against its use. It is true that legumes not coppered will keep indefinitely, but they gradually contract a slight taste of the can, become yellow in cooking, and are little sought for. France has almost exclusive control of the industry.

Such are the principal arguments of the canners.

As regards the consumption of coppered vegetables in the canneries by the workmen, those questioned by the commission denied that they ate the goods to any extent, averring that they soon acquire a distaste or even a repugnance for articles of food which they are called upon to handle so much.

It is true that no accidents are recorded from the use of small quantities of copper. However, its effect is uncertain, and although modern work shows that it is infinitely less dangerous than has been supposed, yet the hygienist can not in the name of science declare it innocuous in every dose, nor pronounce otherwise than "if in doubt, abstain."

As for the rest can any one answer for the ignorance and negligence of the workmen, the indifference of the canners, caprices, temptations, etc.? Has it not just been shown that some introduce salts of copper or leave an excess of sulphate of copper in the liquor which bathes the legumes in the can? The consumer prefers them greened, it is said. These products are a luxury; they are set on the tables of the rich and in hotels and restaurants, where they can be made to pass for the fresh product. That is the secret of their demand. It does not follow that the consumer, even if he has learned to prefer green peas to yellow ones, should prefer them greened by copper. His satisfaction rests upon a deception—to say "green peas" is not to say "peas greened by copper." Greening offers no advantage in the process of preserving. If it preserves the aroma it slightly alters the taste.

Nothing can hinder the greening industry from extending into Alsace-Lorraine, Italy, Greece, Spain, or any other country where the same vegetables are well and cheaply grown. The processes of greening with copper are no longer secret, but familiar in all their details. Only the perfection of the French products may enable a large part of the ancient custom to be retained. It is for the manufacturers who do not use copper to make known by all means and especially by their labels the preference to be given to products prepared without the use of copper salts.

Foreign nations are already aroused upon this point. In Germany, Switzerland and England, analyses are made and prosecutions commenced. Very soon the houses and even the countries which practice the use of copper salts will find themselves in bad repute.

In conclusion, taking into account the quantity of copper existing in the animal organism and in many articles of daily food, sometimes in larger quantities than in copper-greened preserves; considering that recent works seem to show that feeble doses of this metal are not dangerous, but that absolute innocuousness has not been shown for small but oft repeated doses; interested as we are in the industry of preserving foods, an industry whose methods can not be completely transformed in a day; we recommend, without approving the principle of greening with copper, to tolerate temporarily the use of copper up to a certain limit.

This limit is placed at the smallest amount of copper which investigation has shown to be sufficient to impart the desired color. It appears to be 18 mg per kilo of peas or other vegetable taken without juice, or 6 mg per can (demi-boite). The amount is slightly more than is found in farinaceous foods, but less than the amount contained in chocolate.

It is advisable to prosecute canners who use larger proportion of copper, or who use zinc or any other metal.

It seems also advisable to allow this temporary and limited toleration in order to gain time for the investigation of new methods which can be successfully substituted for those in general use to-day.

The introduction of lead solder into alimentary substances preserved in cans made of tin plate has given rise to poisoning. The custom is to be deplored. Manufacturers are urged to substitute other means not involving the use of materials proscribed by law and not constituting serious danger to public health.

The report of this commission was adopted by the Congress.

In 1879 Gautier published a paper[1] which is substantially the same as the report just abstracted.

On March 15, 1878, M. Pasteur reported[2] to the Paris prefect of police the results of the examination of twenty-five samples of canned vegetables submitted to him. Six samples were found to contain copper. Prosecutions were instituted against the vendors of these samples and the matter laid before the procureur of the Republic. He submitted samples of the goods which had been seized to a commission composed of MM. Brouardel, Riche, and Magnier de la Source. They were instructed to report as to the quantity present, to decide whether its presence constituted an adulteration, and if its presence was dangerous to health of the consumer. This committee reported in substance as follows:[3]

Adulteration may be defined to be the addition to an article of food of a foreign substance for the purpose of fraud or gain. Fraud does not exist in this case, inasmuch as the practice is sufficiently well known. The manufacturers whom we visited cheerfully gave all desired information. The fact of the addition has been published by Galippe and Gautier, to the latter of whom it was communicated in an official way by the president of the association of Parisian canners. However, if it be not an adulteration it is certainly a violation of the law.

In spite of the existence of the prohibitive regulations, the packing of copper-greened vegetables is a prosperous industry. Paris and its vicinage pack (yearly) 4,000,000 to 5,000,000 cans (demi-boites), each holding about 300 grams; Nantes and Brittany, 4,000,000; Bordeaux, 4,000,000 to 5,000,000; Angers, Le Mans, etc., 3,000,000 to 4,000,000; and Perigueux, Cahors, Agen, 2,000,000 to 3,000,000.

In regard to the toxicity of copper salts, it may be said that it is almost impossible to take a dose large enough to produce death, both from their horrible taste and from the violent vomiting which they produce. In small quantities the taste is not

---

[1] Des conserves alimentaires reverdis au cuivre. Ann. d'hyg. publ., 1879, [3], 1, 5.
[2] Brouardel, Ann. d'hyg. publ., 1880, [3], 3, 204.
[3] *Loc. cit.*

COPPER-GREENING IN FRANCE.

perceptible, and the salts are not only tolerated but absorbed. Workers in copper are often completely saturated with the metal, but do not suffer from it. Experiments on the animal and human subject have never given a worse result than vomiting or a temporary fit of colic.

Copper normally exists in the human body. It gains entrance from various foods and drinks in the absence of all adulteration. It accumulates to a certain extent, but injury from this accumulation is unknown.

In the samples submitted copper exists to an extent varying between 16 and 45 mg per kilo.

In conclusion, we will state that the amount of copper sulphate in the samples submitted to us does not constitute an adulteration, but that the presence of any copper whatever is illegal. The quantity found by us does not constitute a danger to health.

As a result of this investigation the procureur directed that the suits against the retailers be dropped. A letter to the prefect of police relative to the matter concludes, however, with the following words:

Under the circumstances I submit to you the report of the expert committee, and beg of you to resubmit the question, if you think it advisable, to the Conseil d'hygiène.

In accordance with this request the conseil formed a new committee of three, consisting of MM. Poggiale, Pasteur, and Brouardel. M. Poggiale, however, died before the completion of the committee's labors. The committee[1] submitted a report, of which an abstract follows:

The commission is of the opinion that the Government should not assume the responsibility of allowing the use of copper salts in food unless the public be made aware of the fact. Discussion of the toxic effect of copper salts in this or that dose has been going on for a long time. Carefully conducted experiments have been made to show that copper salts are inoffensive and with apparent success. But all conclusions drawn from these experiments are applicable only to the circumstances under which they were conducted, and go only to show that this or that animal, this or that human being is insensitive to the action of copper salts. Generalizations are dangerous.

Similar answers must be made to the question of the propriety of allowing the use of benzoic acid, salicylic acid, and the like in food.

There is but one way for the Government and French industry honorably to escape responsibility in this respect, and that is to require a frank declaration in each case where foreign substances are added to food. For instance, "Petits pois conservés par tel ou tel ingrédient" (peas preserved with this or that substance).

Manufacturers would be at perfect liberty to attach any explanation they desired in the nature of expert testimony as to the harmlessness of the substance used.

Let the industry defend itself. It can not demand that the Government give it authority for such practices when such a concession would make the Government pronounce authoritatively upon questions of hygiene as yet unsettled by science.

The commission recommends that the Government tolerate the artificial greening of peas upon the condition that on each package shall be legibly printed the name of the agent employed.

The report was adopted by the Conseil.

Evidently, however, this report did not cause universal satisfaction, since on February 2, 1880, the minister of commerce and agriculture

---

[1] Brouardel, rapporteur, Verdissage des conserves alimentaires au moyen des sels de cuivre. Rapport de MM. Pasteur, Poggiale, et Brouardel. Ann. d'hyg. publ., 1880, [3], 3, 193.

once more appealed to the long-suffering Comité consultatif d'hygiène relative to the copper question, transmitting copies of the report of Pasteur and Brouardel to the Conseil d'hygiène et de la salubrité de la Seine, and asking for an expression of opinion on it. He inclosed, further, a letter from the prefect of police to the same general purport. A commission consisting of Wurtz, Gallard, Girard, Brouardel, Chatin, and Rochard was appointed to deliberate on the subject. It was to consider whether it were possible to can leguminous vegetables, producing a salable article, without use of copper salts; in what proportion copper salts were found in commercial goods and in what shape; and, lastly, if such quantities were dangerous to health. It reported, by Gallard, somewhat as follows:[1]

Legumes preserved by Appert's process take a color disagreeable to the eye, and in spite of their excellent quality are not liked in commerce. Greening artificially is almost a necessity.

There are in Paris two canneries which impart to their goods a beautiful green color without the aid of copper. In one of these greening is done with chlorophyl extracted from spinach. In the other the natural green of the peas is retained by a complicated process involving the use of lime sucrate, salt, soda, and sulphite of sodium. Analyses of samples from both factories were made, but in neither case was copper found.

One of the largest of the Parisian canneries employing the copper process was visited. In this establishment 45 grams of copper sulphate are used with every 45 liters of small peas. The resulting peas on examination gave 270 mg of copper per kilo. From this it was calculated that 41.550 grams of the copper sulphate originally employed were absorbed by the peas and 3.450 grams remained dissolved in the water (120 liters) of the coppering bath.

Copper in this great quantity was not found in other goods furnished by this factory. In these the copper descended to 170 or 180 mg, averaging about 175. Analyses of preserves from other factories gave approximately the same figures, being in one case 180 mg and in another 195. In these analyses (Chatin) the copper present in the peas alone was estimated, the surrounding liquor being poured off prior to the analysis. Other analyses, giving the copper present in the total contents of the can (Wurtz), gave 45 to 60 mg of copper. Even after calculating for the water there is a difference between the two sets of figures.

It is claimed that a very small quantity of copper (15 to 20 mg per kilo) is amply sufficient to give the desired hue. It is also said that quantities of copper very much in excess of that necessary to color properly would give an objectionable flavor. Leguminous preserves containing 195 and 270 mg of copper per kilo do not betray it to the taste. Not only is it a fact that the copper in vegetables artificially greened can pass the limit prescribed, but it is a fact that it usually does. It would be interesting to determine by actual experiment just what degree of saturation is possible and how the limit varies with the different vegetables.

As to the nature of the compounds of copper contained in artificially greened vegetables this is a matter of secondary importance. They are principally insoluble albuminates, capable of being changed in the digestive process into soluble salts, readily absorbed.

Copper is not a cumulative poison like lead, and is not violently poisonous even in large doses, its action causing vomiting, colic and possibly diarrhœa of a mild type. There is great doubt as to whether death has ever been caused by copper poisoning. This, however, does not appear sufficient ground to the majority of the

---

[1] Recueil des trav. du Comité consultatif d'hygiène publique, 1881, 11, 362.

commission to cause it to admit the perfectly harmless nature of the copper present in artificially greened vegetables. It is not possible to unreservedly admit the impossibility of mild bodily derangements arising from the long continued consumption of copper. The morbid troubles which all the world knows it may cause have appeared to constitute a danger sufficiently serious to awaken the solicitude of the Government. It is thought that the possibility of an error or accident in the factory, allowing the entrance into the food of sufficient of the metal to injure the consumer, is a good reason for maintaining the proscription.

Since there are at least two processes already known by which the desired green color may be imparted to legumes without the use of copper, the commission regards it as eminently undesirable to remove the prohibition now resting upon the employment of copper and refuses to recommend the toleration, even to the amount of 18 mg per kilo, recommended as a compromise by the Hygienic Congress of 1878.

It is objected that the new methods are insufficient, but the commission has examined numerous samples of goods preserved by them, some of which had been kept in stock for a year, and is able to pronounce them eminently satisfactory.

In order to judge correctly of the efficacy of various processes of preservation it is recommended incidentally by the commission that each package bear the date of packing.

The minority in favor of the toleration of the use of copper salts suggests as a perfectly fair means of settling the question that it be required of each manufacturer to describe his process on the label and leave to the consumer the responsibility of selecting. This, it is thought, however, would amount to a prohibition and would be difficult to enforce.

The commission reports that, after having examined the documents submitted and established the possibility of producing preserves of the color desired without the use of copper, it is of the opinion that it is inadvisable to authorize the use of salts of copper in the preparation of foods.

This conclusion was adopted by the Comité at the meeting of April 21, 1881.

The reports of Brouardel and Pasteur also stirred up the Societé de médicine publique. This body decided to pass upon the subject and to that end formed a commission consisting of Brouardel, Bouley, Decaisne, A. Gautier, A. J. Martin, Napias, Proust, Rochard, E. Trélat, and Galippe, the last being reporter. The report received at the meeting of April 28, 1880, was in substance as follows:[1]

The report of Brouardel and Pasteur recommends that coppering be tolerated if the packages of food be legibly marked with a statement of the fact. Would the administration force a business man not only to reveal the secrets of his factory but stamp them on the goods offered for sale? Such an obligation could not be imposed on the packers. If the process is dangerous to health the Government can suppress it. But in the actual state of science it is impossible to prove that the health of the consumer is exposed to any risk whatever by the greening process as at present carried on.

To advertise the presence of copper would be to ruin the canning industry, since a large part of the public still believes, in spite of expert testimony or in ignorance of it, in the toxic nature of copper and its salts.

A can of peas contains 6 mg of copper. If eaten by three persons each consumes 2 mg of copper, which seems insignificant, since most people do not live upon peas exclusively. Furthermore, copper exists in the preserves in the form of insoluble

---

[1] Ann. d'hyg. publ., 1880, [3], 3, 531.

albuminates and is only slightly assimilable. For twenty-eight years the custom of greening has been followed and there is no recorded accident.

Many foods, notably chocolate, contain copper in larger quantities than greened preserves.

The existence of lead in the tinning and solder of the cans is a much more serious danger, coming, as it does, into contact with the liquid contents.

Since copper exists in the animal and in many useful foods, sometimes in larger quantity than in greened preserves; since an experience of twenty-eight years furnishes no example of accident, and influenced by the interests of the canning industry, it is resolved that there is no reason for interdicting the process of copper greening so long as an established limit is observed.

After a discussion between Drs. Rause, Gautier, and Galippe the society unanimously adopted the report.

The most ardent partisan of the absolute innocuousness of copper and its salts was Dr. L. M. V. Galippe, whose extended researches it can not be denied have shed much light on the toxic nature of copper, as well as furnished most powerful arguments for the interested canners.

In 1875 he published in book form[1] a study of the physiological action of copper salts. He concluded from experiments upon dogs that, apart from the temporary emetic, etc., effect of large doses, these salts were not poisonous.

In 1878 appeared his paper[2] on the use of copper vessels in cookery, in which he endeavored to show that no harm was to be apprehended from such use. His statements may be summed up as follows:

The object of this paper is not to show that cookery performed in copper vessels is superior to any other, but that it is devoid of the dangers usually attributed to it. The prejudice against this metal has gone so far in the past that in Sweden, under Queen Christina, a statue was erected to Prof. Schoffer, who had been largely instrumental in securing a prohibition of the culinary use of copper pots and pans.

For fourteen months I had all the food of my family cooked in copper vessels, and my diet was as varied as possible. No trouble was experienced by anyone, women, children, or visitors. If, however, there was no trouble there were certain inconveniences. Food, especially fatty articles, thus cooked readily turns green. This inconvenience, however, is purely optical. Certain other foods, particularly peas and beans, resting long periods in contact with copper, often absorb enough of the metal to give them a feeble metallic taste, capable of offending a delicate palate. Dr. T. M. Jenkins, United States Commissioner to the exposition (Paris), was of the opinion that the glycerophosphoric acid contained in the yolk of an egg, was capable of forming a poisonous compound with copper. I therefore mixed milk and eggs in a copper pan and heated and stirred till the mixture was of the consistence of cream, and then allowed it to stand twenty-five hours. At the end of this time that part in contact with the pan was green from the action of the air and sour milk on the metal. The general appearance of the mixture was by no means agreeable and its taste very far from being so. Nevertheless it was eaten and with no bad results.

By legal enactment the tin for plating culinary vessels and the like is allowed to contain 5 per cent of lead. From such an alloy food when hot, especially if containing much salt, dissolves lead. Tin also goes into solution. Personally I prefer salts of copper to those of either tin or lead, and therefore prefer copper cooking utensils.

---

[1] Études toxicologique sur le cuivre et ses composés. Paris, 1875.
[2] Ann. d'hyg. publ., 1878, [2], **50**, 426.

On May 20, 1881, M. Tirard, the minister of agriculture and commerce, addressed a circular[1] to the prefects of police, calling their attention to the copper laws. Abstracted, it reads as follows:

At this time, when the preparations for the season's packing of fruits and legumes are under way, I wish to call attention to the fact that the greening of these foods by means of salts or vessels of copper is formally prohibited.

It is important that this prohibition be once more brought to the attention of the canners and retailers. They have lately obtained some opinions favorable to the copper process, but the Comité d'hygiène considering the question anew has decided that it should not be allowed by reason of the danger which it may present to public health.

The Comité was influenced in demanding the continuance of the interdiction of the copper process by the fact that other processes exist which are in successful operation in large factories.

Therefore I invite you to inform the packers and retailers of your department that they expose themselves to risk of prosecution if they green vegetables by means of copper.

M. Tirard issued a further circular on June 28, 1882. It appears that retailers prosecuted under the instructions of the previous circular had set up the successful defense that they were not mentioned in the law of 1860, which prohibited simply the use of copper in greening vegetables, and not the sale of such goods. This circular in substance read as follows:

This interpretation is against the spirit of the law and against the interests of public health and can not be allowed.

You [prefects of police] are therefore instructed to insert after (1) of the old law; "It is forbidden to all retailers or dealers to sell, or place on sale, preserves so prepared."

Probably as a result of the enforcement of the provisions of these circulars, though this is not shown by the accessible records, the subject was once more brought before the Comité,[2] for in 1882 a fresh report was submitted to it. M. Gallard was again reporter. He was evidently getting tired of the ceaseless protests of the canners, for his report was severe and to the point. It was as follows:

Gentlemen, the question of greening preserved legumes is once more before the Comité, and we can not hope for the last time; for, however wise, however just the course which the Government will pursue as a result of our deliberations, it can not fail to injure certain interests, which will as usual unite in alleging persecution and demanding revision of our decisions. The Comité in 1881, as a result of most careful investigation, recommended the renewal of the prohibitions placed upon the coppering process twenty years before, the principal reason given being the discovery of satisfactory and harmless processes for obtaining the same results.

This was not enough for the canners, who have fallen into the habit of using salts of copper and do not wish to change, finding the old process more convenient and somewhat more economical. Two of the members of the trade, desirous of satisfying the requirements of the Government, have made the necessary sacrifices in seeking out a new process by which they can procure without salts of copper the green

---

[1] Recueil des trav. du Comité consultatif d'hygiène publique, 1883, **13**, 431 and 432.
[2] Recueil des trav. du Comité consultatif d'hygiène publique, 1882, **12**, 270.

color which the customers insist upon, although it adds nothing to the nutritive quality of vegetables. The others have preferred to defy the law, alleging that conformity to it would ruin a national industry. This argument, which must necessarily have weight with the statesmen in charge of commerce and industry, is a flagrant error which we can not leave uncorrected. In our previous report (April, 1881) we have indicated processes now in use by manufacturers by which they are able to produce, without employing a particle of copper, legumes of a green color, apparently as well preserved as those containing copper. These canners, whose efforts the Government is desirous of encouraging, are as well worthy of consideration as those who have refused to conform to the law, and it would be as beneficial to the industry of the country to make their process general as it would be to remove the prohibition resting upon copper. The use of copper is not so beneficial as might be supposed to our trade. London merchants selling Parisian coppered preserves have been prosecuted in the police courts. The suspicion placed upon our products, if perpetuated, will work more injury to our foreign commerce than a prohibition which is known to be based upon so important a concern as that of the public health.

From this point of view neither commerce nor industry is especially interested as to whether the prohibition stays or goes. It remains to be seen whether the question is equally indifferent from a hygienic standpoint. The time is past, it is fair to say, when copper, still regarded as a violent poison, was charged with various misdeeds; when one could not suffer from colic or indigestion without blaming an imperfectly tinned copper saucepan; when pathologists described copper colic along with lead colic and regarded the effects of the two metals as analogous and equally dangerous. We know now that poisoning with copper is so difficult to realize that it may be regarded as practically impossible, and that in any case to cause death or even serious illness, extremely large doses are required. It is also established beyond the possibility of question that preparations of copper ingested in small doses cause no injury to health. The case of workers in industries which employ large quantities of copper or its compounds, and of the canners, who almost daily use the greened vegetables, leaves no room for doubt on the latter point.

But on the other hand it is also equally well known that copper salts administered in sufficient doses have a decided physiological action, which may under favorable circumstances become serious. The action is confined to the digestive organs, causes vomiting and diarrhea, and is invariably temporary. Copper could imperil life only when taken in excessive doses often repeated at short intervals. It is absolutely impossible that any dangerous result should follow from the small quantity of copper necessary to green legumes, a dose estimated at from 16 to 20 mg per kilo. However this limit, which is that given by partisans of the absolute innocuousness of copper salts and by the interested canners, can not be relied upon. In the analyses of Galippe the quantity of copper found varied from 14 to 18 mg per kilo, but Riche and Magnier de la Source found as little as 16 mg for petits pois and as high as 35, 40, and 45 mg in green beans. According to Pasteur the quantity is from 80 to 100 mg of copper per kilo. I myself received a sample prepared as follows:

Forty-five liters of small peas were plunged into about 120 liters of boiling water, to which had been added 45 grams of copper sulphate. I submitted this sample to M. Chatin, who, having analyzed it with the assistance of M. Personne, declared that it contained the enormous proportion of 270 mg of copper per kilo of peas taken without the juice. Chatin persisted in his statement, although it was doubted by some members of the commission. I should add that using M. Chatin's figures for the copper remaining in the coppering bath and the 270 mg number, the entire amount of copper sulphate used can be recovered by calculation. In other samples from the same house furnishing the one just cited, the proportion was not so great, falling as low as 170 or 180 mg per kilo, amounts, however, still much in excess of that reported by Galippe and by the canners in their statement to the minister. In the samples from the same source, analyzed in the laboratory of Wurtz, the amounts fall still lower, but even these reached 78 to 80 mg per kilo. On miscellaneous samples sold under various names, the results ranged from 175 to 195 mg. Results obtained with the permission of the prefect of police from the municipal laboratory of the city of Paris varied from nothing or a mere trace, to 184 mg per kilo. The extreme limit was found in a sample of peas (petits pois) sent in, while in the samples seized by the police the largest amounts were 36 mg in "haricots flageolets," 128 mg in "haricots verts," and 14 in "petits pois." The differences may have been due to the degree of maturity or to the thickness of the cortical envelope of the legume.

At any rate, the diversity of the figures obtained shows that it is quite possible to introduce into the legumes much larger proportions of the metal than are necessary for greening. This proportion being always given as from 16 to 20 mg, it is easily seen that it is habitually exceeded. It has been repeatedly asserted that if the limit were overstepped to any extent the vegetables could not be used, the copper revealing itself to the taste. The reporter has not altered his opinion on this subject, expressed in 1881, for he himself has frequently eaten and had others eat unawares preserves containing anywhere from 180 to 270 mg of copper, and no objectionable flavor was detected. These two points may be considered perfectly established: First, that the limit of 16 to 20 mg of copper is habitually passed; second, that vegetables are capable of absorbing a quantity of copper sufficient to affect the health of the consumer without betraying the fact to the taste. Although most of the commission of 1881 recognized the innocuousness of copper in the quantities in which it is found in preserved vegetables, the majority thought that in certain exceptional cases it could become injurious, and that since it is not indispensable to the manufacture of greened vegetables, it was not desirable to repeal the original prohibition, judging it unwise to open the door to abuses which would not fail to follow were any toleration granted. There is a simple method of reconciling the conflicting interests, affording the interested canners an opportunity to place their goods before the public and leave there the responsibility of choice. This measure, originally the suggestion of Pasteur and the recommendation of two previous commissions, is simply to require each canner to inscribe on his label in distinct characters the nature of the substances added to the vegetable in the process of preservation. Numerous objections have been made to this proposition. It has been claimed that it would be equivalent to a prohibition, since the public in its ignorance would never consent to buy goods bearing this label, but it was rejected principally because in the present state of our legislation it is impossible to force the canners to comply with it. The question has not been allowed to rest by those who are fearful lest their business suffer if the public is to become aware of the presence of copper in their goods.

The association of canners at Paris, that at Bordeaux, and that of Brittany and the Vendée have appealed to the minister for authority to continue the use of copper. The process is to be considered not only from a hygienic standpoint as rendering the product injurious to health, but from the commercial side as constituting an adulteration by which goods are given a color which they do not naturally have, and which the simple process of preservation would not impart. Therefore it is an adulteration and should be prosecuted as such by the authorities.

If the prefect of police has the right to demand that margarine and its products be marked as such, and if the Comité at its last session saw fit to advise the measure, why has not the Government the right and power to pass an analogous measure in the case of legumes prepared by the copper process? If it has not the right there is still the resource of proposing to Parliament the adoption of a law that will give them the right. It would be desirable not only for the present case but in all cases in which industry considers it indispensable to introduce into food—either liquid or solid—a foreign product which does not figure in it naturally, either for the purpose of improving it or of facilitating its preservation. The object is not merely to preserve the public health, but above all to secure good faith in commercial transactions by preventing manufacturers from altering in any way the nature of an alimentary substance without due notification of the fact. A law like this would not have to fix, as has been demanded, the limit where a certain substance becomes dangerous and is therefore to be absolutely interdicted—a point extremely difficult to establish except in the case of undoubted poisons. It is understood of course that such a law would not be applicable in the case of substances whose poisonous nature is known beyond the possibility of doubt—for example, lead, arsenic or mercury, which must always be entirely prohibited, but only to those whose toxic effects are sufficiently mild to admit of doubt.

It is not possible to impose upon us the alternative—"if it is poisonous, prohibit it: if not poisonous, allow unreserved toleration"—for we have previously established that it is not usually poisonous, but may in exceptional cases become so, which, although it does not authorize absolute prohibition, certainly justifies warning. This, it seems to us, would be adequately furnished by the label. It is claimed that it would be impossible to secure the conviction of offenders. This is a profound mistake. On the contrary, magistrates would find themselves much relieved by the removal of the necessity now existing of arbitrarily deciding as to whether the substance added to the food was of a dangerous nature, but having established the presence of the substance and the fact that the label bore no announcement of its presence, decision would be as simple as in the case of the dealer who has added water to milk or wine. Accidents are disposed of. If a labeled can contains enough of the substance to injure the health of the consumer, the canner should be liable to prosecution exactly as is the pharmacist who substitutes 50 mg of an emetic for 5 mg.

This law would be useful. It protects all interests, those of public health as well as those of commerce and industry. It is desirable that it be made for the present the subject of study and presented with as little delay as possible for the consideration of Parliament. Further it would be well, indeed indispensable to this law, that it require the label to give, along with the substances added in preservation, the year in which the preservation took place. This was suggested in 1881:

"To permit a correct judgment of the comparative efficacy of the various processes of preservation it would be well that each package of preserves should bear the date of its preparation. This would be a salutary measure, and incidentally we recommend it."

This view, advanced somewhat cautiously, was taken up by the Conseil d'hygiene et de la salubrité d'Alger in a deliberation which has been sent to the Comité for examination. We can only approve, for the processes of preservation, however perfect, are not indefinite, and the public is exposed to deception in buying hermetically sealed boxes of foods which on opening prove to be entirely spoiled. This is not infrequent, especially in the case of meat and fish. It is true that if the measure be adopted the buyers will not fail to select the freshest products and the canners will not fail to denounce the Comité and to charge them with injuring the national industry in throwing back upon their hands old or damaged stock. But the consumers will gain, and this should be the consideration of the Government to whom is intrusted the welfare and health of the people.

The green color desired by the consumer, although it does not enhance the nutritive quality, can be obtained without the use of copper salts.

The quantity of copper strictly necessary to impart the coloration is so small that it can not cause danger or injury to health, but if by carelessness or mistake the quantity is materially increased, it can cause serious though temporary trouble in the health of the individuals who consume the product thus prepared.

The commission thinks that the possibility of such accident is sufficient reason for maintaining the prohibition on salts of copper.

The public would be sufficiently protected from these accidents if it were possible to draw attention to the presence of copper by imposing upon canners the obligation of announcing by means of labels in legible characters the presence of the metal. For these reasons we propose that the Comité reply to the minister:

Since present legislation does not confer upon the administration sufficient authority to permit it to require the canners to announce by labels any substances added by them to the preserved foods, whatever their nature, it is necessary both in the interests of hygiene and of commercial honesty to demand of Parliament the enaction of a law requiring merchants and canners to place upon their labels an announcement of any such addition, and further to accompany this with the date when the preserves were prepared.

It is not necessary to remove the prohibition resting upon copper before the passing of this or any other law tending toward the same end.

After discussion the Comité presented the following report:

In the actual state of science it is not shown that the greening of preserved foods by salts of copper is absolutely without injurious effect, and therefore it is not advisable to remove the prohibition.

Decided as was Gallard's second report it was by no means accepted as final.

In February, 1886, the association of Parisian canners (Chambre syndicale des fabricants de conserves alimentaires de Paris) complained to the minister of commerce and industry of being hampered in their trade by the copper laws, stating that the buyers in the United States had advised them that their importations must be diminished were the canned vegetables shipped no longer coppered, for American commerce demanded the regreened legumes. It was also stated that London merchants had written them to the same effect, stating that vegetables of the natural color would not be acceptable. Furthermore the canners stated:[1]

We do not know any practical and approved method of giving leguminous preserves the green tint demanded in trade, especially in foreign trade, except that depending upon the addition of an infinitesimal dose of copper salts during preserving. A trade of fifty years' standing demonstrates the perfect innocuousness of the amount employed. The Comité d'hygiène recognized in 1882 that copper sulphate in the quantities we use is harmless, but still maintained the prohibition on the ground that the manufacturer might use more copper than was necessary, and thus cause illness. In short the whole industry is to remain under suspicion because some bungler may not know his trade. Attention is invited to the fact that the manufacture of copper-greened preserves occupies 20,000 workmen and represents 40,000,000 francs. Enforcing the prohibition would send this trade to foreign countries.

They close by saying that if salts of copper are as poisonous as has been claimed, the packers should be prosecuted in real earnest and thus given a chance to demonstrate in the courts the innocuousness of their products. They say in conclusion:

"If we are criminals, let us be judged seriously."

In May, 1887, the chamber of commerce of Nantes protested against the partiality with which the law was enforced, Parisian canners not being prosecuted. It demanded that if the proscription was to be continued, it be rigorously and impartially enforced.

---

[1] Reverdissage des conserves alimentaires au moyen des sels de cuivre. Ed. Grimaux, rapporteur. Recueil des trav. du Comité consultatif d'hygiène publique 1889, 19, 146.

The prefect of the Loire-Inférieure stated in connection with this protest that in his department several canners who strictly observed the ordinance were at a great disadvantage in competing with their rivals. The prefect said farther that the interdiction injured the interests of French trade, and requested that an examination be made into the matter as to whether the salts of copper were really dangerous, and requested that pending the results of the investigation the employment of small amounts of copper sulphate be tolerated.

In accordance with these requests a new committee was formed with Grimaux as chairman. Grimaux reported the results of the commission's labor to the Comité consultatif on April 15, 1889. Briefed, it was as follows:

In 1860 salts of copper were still regarded as poisonous. Subsequent investigations have certainly modified this view.

As for the report of my colleague, Gallard, in 1882, it does not appear to be justified, for his arguments are rather on the other side, his conclusions being based on the statement that other successful processes exist, which is contradicted by Bouchardat and Gautier. If there exists no other argument against the interdiction than that no accident has resulted from the use of the copper-greening process and that no other greening process exists, it seems that the Comité would assume grave responsibility in continuing the prohibition.

Is it advisable to label these vegetables as suggested by Brouardel and Pasteur "légumes au naturel" (natural legumes) in the one case and "légumes reverdis au cuivre" (legumes greened by copper) in the other? This was not accepted by the Comité because it amounted to an interdict, as the public, not being informed of the harmless nature of copper, would reject these goods and foreign markets would be closed.

It has been proposed to mark in a special way packages intended for exportation and to limit the prohibition of salts of copper to those goods intended for home consumption, but this seems impracticable, and would furthermore expose the French goods to restrictions in foreign markets. There remain but two alternatives, absolute freedom or a prescribed limitation of the quantity of copper to be used.

The duty of the hygienist is difficult in this matter. On the one hand his legitimate tendency is to proscribe the introduction into alimentary substances of all bodies which are not food and which are toxic in any degree, but in taking this as a rule we may put restrictions upon our industries, and this is a consideration which the interests of our country can not afford to ignore. The fact that copper is found in the legumes in the state of insoluble albuminate leads us to conclude that there is no danger to public health in authorizing without restraint the use of the copper-greening process, but since the purchaser has a right to choose his food, it is deemed advisable to label the products of the two processes respectively, "au naturel" and "à l'anglais" (after the English manner). These are the conclusions which the reporter submits. One member of the commission, however, raised the point that as copper had some antiseptic action, it must have, even though not poisonous, some retarding influence upon digestion.

Experiments were made upon this point by M. Gley. In one experiment 10 grams of haricots verts (green beans) colored by the copper process were cooked and ground with 10 grams of blood fibrin, and after covering with 60cc of gastric juice, were kept at 45° for 48 hours. The fibrin was entirely dissolved, but the beans were not attacked. This was because they are largely cellulose. In the second experiment 2 grams of copper sulphate dissolved in 40cc of distilled water were added to 60 cc of gastric juice in which were placed 20 grams of blood fibrin, and the whole heated to 45° for eighteen hours. At the end of this time the fibrin was entirely dissolved. These experiments did not allay the scruples of our colleague, who claimed justly that the experiments did not settle the question of the digestibility of the vegetables in question. A new set of experiments was then made with great care by M. Ogier to determine the comparative digestibility of vegetables greened with copper as compared with those not so treated. M. Ogier concluded that there is nothing to indicate that the digestibility of greened vegetables is less than that of those "au naturel."

Among the facts furnished by our work are these:

That the quantity of metallic copper contained in green legumes of commerce is about 13 mg per kilo; that with 10 mg they are sufficiently colored, and that 40 mg impart a taste which renders them inedible.

We propose that the Comité consultatif reply to the minister that it is of the opinion that in view of the present state of our knowledge of the toxic effects of the salts of copper, it is not necessary to interdict the process of greening with salts of copper.

The report was approved by the Comité consultatif d'hygiène on April 15, 1889.

On April 18 a ministerial circular was issued, raising the prohibition on the use of copper-greening for preserved foods, directed to the various prefects. It read as follows:

Three ministerial circulars—(1) December 20, 1860; (2) May 20, 1881; (3) June 20, 1882—have been addressed to the prefects instructing them to take action to prevent manufacturers, merchants, and retailers from employing vessels or salts of copper in the preparation of preserves of fruit and legumes destined for alimentation; also to prevent selling or offering for sale goods thus prepared.

This prohibition was based upon the repeated statement of the Comité consultatif d'hygiene publique of the danger of poisoning from the introduction of copper sulphate in frequently repeated doses into the human system.

The action in such cases of salts of copper upon health had not at that time received the careful study which has since been given to the subject.

In the last few years new studies and extended experiments which have been now brought to a successful close, have resulted in the establishment of facts hitherto doubtful.

The question having been submitted anew to the Comité consultatif d'hygiene publique, it has also conducted experiments to determine the dose which can be safely ingested and the proportion contained in copper-greened vegetables.

From this it has been concluded that the recent knowledge acquired was of such a convincing nature as to justify them in no longer opposing copper greening.

I have the honor, therefore, to call your attention to this conclusion.

In order to promote the interests of commerce and to render to it the satisfaction so long denied, it is for you to take action conformable to this circular. Let the necessary measures be taken without delay, and let information of the execution of the order be transmitted.

Nothing is said in this circular relative to the distinction between the two classes of goods which was proposed by Grimaux—possibly because it would not look well upon the records. The matter may have been left to the discretion of the prefects. In the samples of canned vegetables examined in this laboratory, those French goods marked "au naturel" always contained copper, and though those marked "à l'anglaise" usually did so also, there was one which did not, the pea sample No. 10629. This sample, however, contained a large amount of zinc, as a compensation.

The raising of the prohibition was done with much haste as is evinced by the close correspondence between the dates of the report of the commission and the date of the ministerial decree. The proposition of M. Grimaux to mark the distinction between coppered vegetables and those not so treated, by a meaningless combination of words like "à l'anglaise," meaningless, that is, in this connection, can scarcely be viewed as a happy one, taken from an ethical standpoint.

The statement of the canners and others, that if coppered vegetables bore a label announcing that fact in so many words, no sale could be effected, was naive. The whole struggle in France over the copper question may be summed up in a few words. Certain packers of food products wanted to use in preparing their goods a substance which, rightly or wrongly, was viewed by the consumers of these foods with distrust, and, being aware of this fact, petitioned for permission to add this substance without the knowledge of their customers. The Government, acting upon the counsel of its hygienic advisers refused this permission for many years, but finally conceded it, acting, as it appears, altogether upon commercial considerations. The report of the scientific commission upon which this license was nominally based is also actuated by this motive, for M. Grimaux states this fact plainly, stating that if it were not for economic considerations his duty as a hygienist would lead him to another conclusion. His recommendation that the goods be marked when coppered, though with what is practically an arbitrary sign, does not appear in the ministerial circular, nor does any limiting amount, so that as far as the records accessible show, the French packer is at full liberty to add as much copper as he pleases and without advertising the fact in any way.

COPPER-GREENING IN BELGIUM.

In 1883 Belgian packers were prosecuted for coppering vegetables.[1] The decision was in favor of the defendants, being in substance that they had not added substances of an injurious nature to the articles of food in question. On November 17, 1885, the Belgian minister of the interior addressed a letter to the Royal Academy of Medicine of Belgium, asking for a decision as to whether salts of copper occurring in food were injurious. After an exhaustive discussion of the subject the academy came to the conclusion that "compounds of copper are not only useless in foods, but they are injurious."

COPPER-GREENING IN GERMANY.

In their anxiety to obtain new outlets for their traffic, Germans have long looked with covetous eyes on the great export trade in preserved vegetables possessed by France, and the repeal of the clause in their

---

[1] Cited by W. H. Kent in a paper "On the use of copper in foods," report of Brooklyn Board of Health, 1887, 135.

COPPER-GREENING IN GERMANY.

food law by which copper-greening is prohibited, has been vigorously agitated for some years. In 1890, the Freie Vereinigung bayerischer Vertreter der angewandten Chemie (Independent association of Bavarian representatives of applied chemistry) resolved to investigate the subject. At the Augsburg meeting, July 18, 1891, Mayrhofer reported in substance as follows:[1]

Preserves cooked in copper vessels, but not intentionally colored, contain small amounts of copper. In the following table the figures given represent milligrams of copper found in a kilo of the food (parts per million):

Name.	Copper.	Name.	Copper.
Hazelnuts	3.1	Cherries	2.3
Apricots	1.0	Medlars	2.8
Gooseberries	4.2	Currants	8.0
Peaches	5.0	Pears	4.2
Raspberries	4.2	Strawberries	8.0

Samples of green-colored fruits were found to contain:

Name.	Copper.	Name.	Copper.
Figs	15.1	Almonds	36.8
Do	19.0	Do	22.8
Do	17.0	Do	26.5
Chinois, green	47.0	Chinois, yellow	56.1
Do	76.6	Greengages	18.0

Raw, uncolored fruit contained:

Name.	Copper.	Name.	Copper.
Chinois, green	1.1	Chinois, yellow	0.9

In the case of green vegetables there were found:

Name.	Copper.	Name.	Copper.
Cucumbers	45.0	Green peas	44.8
Peas	25.0	Do	40.0

In the reporter's opinion presence of copper to the amount of 20 to 24 mg of copper per kilo of vegetables is sufficient to impart the desired green color, and he proposes that this amount be declared allowable.

Reference was made to the fact that if coppering canned foods were prohibited the German industry would be handicapped in competition with the French.

DISCUSSION.

In the subsequent discussion Dr. Karsch remarked that he understood that the French limit of 24 mg had been adopted as a result of a case of poisoning with cucumbers.

[1] Ber. bayr. Vertr. angew. Chem., 1891, **10**, 77.

Barth recalled the Brunswick trial, and remarked that the accused in this case had been acquitted, as in the opinion of the court he had not used copper to conceal the condition of damaged goods, and also the amount of copper present and the form in which it occurred were not injurious. He hoped that the society would declare that in its opinion the use of small amounts of copper to color vegetables should not be condemned. It is a positive fact that for this class of goods some kind of coloring agent is a necessity. Vegetables, especially peas and string beans, retain their natural green color on simple cooking in open vessels, but when heated in the sealed can, become yellowish or brownish and are unseemly in color. Peas can be tolerably easily colored with various vegetable colors, but this is not the case with beans. Copper salts are administered as emetics in doses as high as a gram, and he had seen the fatal dose given as 10 grams. These, however, were amounts which were far in excess of anything which could possibly be taken with canned goods. To his knowledge chronic copper poisoning was unknown. Cows had been given doses as high as 8 grams of copper sulphate daily without bad result.

Dr. G. Merkel agreed with the previous speaker that acute poisoning from large amounts of copper could not occur from the use of copper salts for coloring canned vegetables. Chronic copper poisoning had not been proved. The cases in the literature could not stand a severe criticism. He himself had lately given some attention to the subject, and had come to the conclusion that the state of knowledge of the toxicology of copper is defective. In the course of a long experience as a hospital physician he had observed no case of chronic copper poisoning, although if it occurred at all it should not be rare. In the trades having to do with copper chronic copper poisoning has not yet been observed. On the other hand, it is not yet possible to say that the metal is harmless. The state of the copper occurring in the different canned foods has a bearing on the question. It is quite possible that small doses of copper if repeated for some time might, under some circumstances, prove dangerous, but he was not prepared to state what these circumstances might be. It was dangerous to attempt to draw conclusions in regard to the human subject from the results of experiments upon the animal. He was opposed to applying the pharmacopœia to food chemistry. At present he did not regard the question as decided, and was of the opinion that medical men should wait for further results before committing themselves.

Kayser spoke of the danger of rushing from one extreme, according to which copper salts have been regarded as dangerous, to the other and calling them harmless. Preserved foods are not merely an occasional article of diet, but in many instances form an exclusive food for considerable periods of time. He was of the opinion that it was not yet time to form an opinion.

Dr. Egger said that at present it could not be said definitely what amount of copper was dangerous. Much depends upon the individual. He was of the opinion that it would be a good thing to establish a maximum limit.

Mayrhofer remarked that the question was not one of soluble copper salts, to which Kayser retorted that the question involved such solvents as gastric juice and not water or dilute hydrochloric acid, and inquired if Mayrhofer and the canners were ready to guarantee that the 20-24 mg limit should not be overstepped.

Mayrhofer answered that in making some dyeing experiments with peas two samples were colored to nearly the same shade, and absorbed, respectively, 18 and 24 mg. In two other experiments the coloring was pushed till the peas had a marked blue shade. The amounts absorbed were, respectively, 32 and 44 mg. Coloring in this way (with copper salts), the risk of such amounts as he had found in the chmois sample coming into the goods was quite excluded. On inquiry into this case he had found that the large amount (76 mg) was due to the fact that the fruit had been allowed to stand several days in a copper pan. This, however, was an accident and such cases must be rare. In his coloring experiments he had found the minimum amount of copper with which the desired color could be obtained to be 17 mg.

Hilger recalled the fact that in the case of these vegetables the copper entered the body as an insoluble salt. When copper salts are added to albuminous liquids, even though these be slightly acid, all copper is precipitated as an insoluble compound, and for this reason copper when present in food is found in combination with the albuminoids. He believed that copper was a normal constituent of the human body. He was in favor of postponing action upon the question to the next yearly meeting.

Dr. Koehler remarked that in fixing maximum amounts the pharmacopœia could not be safely taken as a guide. He also stated that at the Strasburg exposition he had seen some beautifully-colored green beans which derived their color from being cooked in a copper kettle, an electric current being simultaneously passed through them, the kettle serving as anode. Naturally large quantities of copper went into solution.[1]

It was finally resolved that the meeting express no definite opinion, but that the whole subject be postponed until the next yearly meeting. At the 1892 meeting, held at Regensburg August 2, Lehmann reported on the copper question. Abstracted, his report reads:[2]

The method used for the determination of copper was to moisten the organic matter under examination with sulphuric acid (occasionally with nitric), slowly dry, burn in porcelain crucibles, leach the ash, and burn the carbon with soda and saltpeter. The ashes were dissolved in dilute hydrochloric acid and precipitated with hydrogen sulphid in weakly acid solution. The filtrate from the precipitate was again treated with hydrogen sulphid. The sulphids, together with the filter on which they were collected, were burned and the ash dissolved in hydrochloric acid and treated with ammonia. The copper in the solution obtained was determined in cases where it amounted to more than 2 mg by the method of De Haen (sodium thiosulphate and potassium iodid). Amounts of copper lying between 2 mg and 0.3 mg were determined colorimetrically with ammonia. Less quantities were colorimetrically determined by ferrocyanid. Mach's hydrocyanic acid and gum guiacum method was not found advantageous, although it allowed the estimation of still smaller amounts. Numerous trials with known amounts of copper were made as a control, and nearly all tests were made in duplicate where the amount of material made this possible. All reagents were copper-free.

The results of Mayrhofer were quoted and commended.

Determinations of the amount of copper in the livers and kidneys of various animals were made. The results were:

Name.	Copper.	Name.	Copper.
Sheep liver	18	Calf liver	48
Ox liver	51	Dog and cat livers	10–12
Ox, sheep, and rabbit kidneys	3.8–8.0		

French authors were quoted to show that sea animals often contain larger amounts. A dozen oysters are stated to contain an amount of copper lying anywhere between 36 and 108 mg, being a maximum of 2000 mg per kilo.

As to the amount of copper present in artificially colored vegetables, Mayrhofer's results were stated and also results drawn from other sources. The highest amount on record was given as being between 184 and 270 mg per kilo. The reporter had tasted goods containing 244 mg, but could not detect copper. The taste of the peas was wholly unaltered. Peas coppered by himself and containing 634 mg of copper per

---

[1] See page 1162.
[2] Ber. bayr. Vertr. angew. Chem., 1892, 11, 16.

kilo were eaten to the amount of 200 grams in one sitting and without any accompanying article of food. There was no unpleasant taste at the time.

In uncoppered vegetable foods amounts of metal up to 30 mg may be found, and in animal foods up to 50. In the case of vegetables and fruits carefully colored, the copper content ranges between 25 and 50 mg, but when the operation is conducted with gross carelessness the amount may ascend to 270 mg without being noticeable.

In regard to the form in which naturally occurring and artificially added copper occurs, little is known upon this point. In the case of vegetables, Tschirch has proved that at least a part of the added copper is united with chlorophyl as copper phyllocyanate. However, as peas contain but little chorophyl this amount reaching, according to the same author, 0.05 per cent of the dry matter, it is easily seen that only a small part of the metal can be combined in this way. Peas contain 80 per cent water, so that a kilo of fresh peas contains but 100 mg of chlorophyl, which could unite with only 10 mg copper. It is probable that the rest of the copper is present in the form of albuminate. Some may occur as phosphate. Experiments on the solubility of the copper in strongly colored peas in dilute hydrochloric acid showed that only a small portion dissolved. Addition of pepsin did not affect the solubility.

Peas should not be harmful if eaten in quantities of a pound (500 grams). Such a quantity as this is seldom eaten at a meal, so that this may be put as an outside limit. Peas containing 25 mg per kilo must be absolutely harmless, inasmuch as, judging from experiments with artificial gastric juice, but 1 or 2 mg could be dissolved in the stomach. The innocuousness of these small amounts has been clearly demonstrated by time, for French peas have been coppered for years. No case of poisoning from their use is on record.

The reporter instituted experiments both on himself and on a friend. Amounts of copper sulphate equivalent to 75, 120, and 127 mg of copper were mixed with peas or beans, and the vegetables cooked in enameled pots and eaten in two portions, one at noon and the other in the evening. The experiment was not exactly pleasant. The vegetables gave no unusual taste at first, but the peculiar copper taste soon appeared and the food became unendurably repulsive. On waiting an hour or so, however, a fresh portion could be eaten. In the before-mentioned case, in which 200 grams of peas containing 127 mg of copper were eaten at one time, unaccompanied by other food, no particular sense of repulsion was experienced. Subsequently, however, an intensely disagreeable copper taste developed itself. He had endeavored to overcome this with a little wine. An hour after the meal a sense of illness developed and lasted two and a half hours, relieved finally by two violent attacks of vomiting. In the evening appetite was fully restored.

Therefore amounts of copper exceeding 100 mg are capable of exciting mild disturbances of health, a bad taste, vomiting, etc., but nothing more. Colic or slight diarrhea is also possible, but it was not observed.

Different animals were given amounts of copper ranging between 10 and 100 mg per day, daily for months, without noticeable effect. The author himself and his friend took a daily dose of copper for a month or so, starting in with 20 mg and increasing it to 30, or double the amount which would be found in 500 grams of preserves. The action was absolutely nothing. The metal was taken in beer, partly in the form of acetate, partly as sulphate.

These experiments do not consort with the cases of poisoning reported in the literature. In these, as a general thing, food has been allowed to stand for some time in copper vessels. Experts have found copper in the food or in the body, and the toxic symptoms have, without further investigation, been ascribed to this metal. In the opinion of the reporter these cases were due to action of ptomaines.

Speaking generally, it may be said that from a hygienic standpoint all unnecessary additions to food should be discouraged. The only benefit of the coppering process, inasmuch as taste, nutritive value, and digestibility remain the same, is the improved appearance of the goods. No disadvantage is found in the addition of the copper as long as the addition is a small one. If the amount be larger slight bad results may be produced, but very bad results are not to be feared even with very careless work. In the opinion of the reporter 25 mg per kilo are insignificant from a hygienic point of view, but 100 mg are not altogether a matter of indifference. A line at which injury to health begins, however, can not be drawn.

In making a final conclusion upon this point purely practical questions should be considered, and it must be left to the technical expert to decide whether the commercial advantages of allowing the use of this substance overbalance the possible danger of the goods suffering discredit in the minds of those ignorant of the subject. It is, however, best to fix a low limit like 25 mg. It may be finally reiterated that hygienically there is nothing against the concession.

DISCUSSION.

In the subsequent discussion Mayrhofer spoke of the results of Tschirch. He further remarked that there was an opinion that copper was unnecessary, inasmuch as there were other non-injurious colors which could serve the same purpose. This was true for many substances, but not for canned foods. A coloring matter has been sold in Vienna for this purpose, but it did not resist the action of acids.

Hilger remarked that in some investigations on marine animals he had frequently discovered copper. In some of the Tunicatæ he had discovered 0.02 per cent copper (200 mg per kilo). In salt and fresh water crabs he had also detected it. Copper is a normal constituent of the human body. In the neighborhood of Treuchtlingen a a copper-poisoning case occurred. The patient ate some soup and died in a few hours. The soup contained a large amount of copper. A fatty soup standing in clean scoured copper vessels can absorb considerable metal. In one case this amounted to 0.163 per cent copper.

Kayser remarked that, in discussing the presence of copper in preserved foods, it was irrelevant whether it was in part of natural origin or all artificially added. He was of the opinion that the limit should be placed at 25 mg and that this limit should not be overstepped.

Borgmann mentioned the danger of copper contamination from Bunsen burners in carrying out tests for the metal.

Lehmann remarked that he had been on the watch for this danger, but had found no evil results from the ordinary brass Bunsen burner.

Von Kerschensteiner said that it had long been known to physicians that copper was a normal constituent of the body, either in small amounts or in larger accumulations. It is also a piece of familiar knowledge that occasionally copper exerts a poisonous action. It was, however, now evident that copper salts administered up to a certain amount were not injurious, and in the future cases of illness resulting from food which was found to contain copper would be regarded in a different light. It is easily conceivable, however, that a small dose of copper administered in the form of one organic combination might have a different action from that of the same dose in another shape. Maximal allowable limits might be different for different substances.

Kümmerer inquired if the copper occasionally present in mineral waters would be likely to exert a different action from that present in canned foods.

Halenke thought caution and further knowledge were necessary in taking a stand on the matter of maximum limits, in view of the great fear which had hitherto been entertained of copper.

Kayser was of the opinion that the time had come to take a definite stand in the matter.

Lehmann remarked that time was expensive, and that since 1818 30 or 40 authors had handled the subject. Cuprophobia dated from the middle of the previous century, but he thought it was without foundation. As to the impression that one copper salt might have different action from another, he would say that he and previous investigators had tried many salts, chlorid, sulphate, nitrate, acetate, butyrate, oleate, stearate, succinate, albuminate and others, and upon the whole the results agreed very well. He also stated that at one time Galippe had taken the trouble, to convince an American who cherished a conviction that copper glycerophosphate was specially injurious, of cooking a complicated mixture of eggs and milk in copper vessels, allowing to stand and finally eating the unappetizing green mass. No injury had resulted. As to the idea that there might be a coöperative action between the products of putrefaction and copper salts, this had been expressed by Dr. Pasch in 1850. In his opinion, this was not impossible. In the literature there were cases on record which were difficult of explanation without assuming an occasional abnormal toxicity for copper salts. These isolated and difficultly explainable cases, however, should not deter from taking profit by the results of exact laboratory experience. He reiterated his conviction that there were no hygienic grounds against the toleration of 25 mg of copper to the kilo of preserves.

Sendtner recalled the fact that the Italian Government had declared 100 mg allowable.

Halenke remarked, after the further explanation by Lehmann, that he thought the limit could be allowed, but only for preserved foods.

Hilger said that it was time to settle the matter.

Hörman thought that it was of economic importance for the canning industry that the question be finally settled.

A discussion now followed as to the exact form the resolution fixing a maximum limit should take. It was closed by the presiding officer calling upon Mayrhofer to present the resolution. Mayrhofer moved:

"Judging by experience an amount of copper in preserved foods not exceeding 25 milligrams per kilogram is not to be viewed as injurious to health."

The resolution was unanimously passed.

In how far the views of the Bavarian chemists were biased on this point by their very evident desire to promote German commerce is a question hard to decide this side of the water. At all events, their reports give the use of copper in foods a decided "whitewashing."

The Brunswick case alluded to occurred in 1891.[1] A Brunswick preserving firm used copper sulphate to green peas, and the authorities instituted legal proceedings against it. In the preliminary proceedings, among other testimony it was stated that an adult would consume at a meal about 170 grams of peas, corresponding, in the case of the goods in question, to about 6.5 mg of copper at the highest, and that in this quantity copper as copper sulphate would not be at all dangerous, even in repeated doses. It was also stated that physicians frequently administer copper sulphate as an emetic in doses of 200 mg, and that the pharmacopœia gave a gram as the maximum dose. As a result of this evidence the defendant was acquitted, the court stating that he had but exercised the ordinary mercantile right of beautifying (herausputzen) his goods.

COPPER-GREENING IN ITALY.

In 1892 the Italian Government[2] amended the food laws so that the section which had hitherto read that preserved goods containing more

---

[1] Chem. Ztg., 1891, 15, 49.   [2] Ztschr. Nahr. Hyg., 1892, 6, 269.

than 100 mg of copper per kilo were to be condemned, now reads that amounts of copper not exceeding 100 mg per kilo are to be allowed in green vegetable preserves.

COPPER-GREENING IN GREAT BRITAIN.

In England the practice of greening foods with copper has never been favorably regarded. In Accum's celebrated book "A Treatise on Adulteration of Food and Culinary Poisons, exhibiting the fraudulent adulteration of bread, beer, etc.," better known by its startling subtitle, "There is Death in the Pot," much attention was paid to the subject. It seems that the cook books then in vogue recommended cooking pickles and vegetables with half pence to insure their retaining their green color. This practice Accum strongly objected to. A great part of his book indeed was devoted to the contamination of food by copper, and he cites many cases of poisoning. He states on page 353 (second edition, London, 1820) that the senate of Sweden in 1753 prohibited copper culinary vessels, and ordered that none but such as were made of iron should be used in the Swedish fleet or armies.

In 1851 the London Lancet founded a commission to investigate food adulteration in London, placing it in charge of A. H. Hassall. The investigation continued through the next three years, and the results were published in book form in 1855. Hassall found that the use of copper for greening vegetables was quite common. Ten samples of mixed pickles, 4 of gherkins, and 3 of pickled beans all gave good copper tests. Hassall characterizes these results as "simply fearful." He also found that 27 out of 34 samples of bottled fruits and vegetables contained more or less copper. With fruit preserves, marmalades, etc., 33 out of 35 showed the occurrence of copper.

In almost all of the published prosecutions of English dealers for selling coppered peas convictions have been obtained. Public opinion appears to be strongly against the practice.

In 1877 and 1878 much attention was paid to this matter The Analyst for those years contains many articles on the subject.

In 1890 the city authorities of Glasgow appointed a committee to investigate the subject, and their report has been published in pamphlet form under the title, "Report on the Greening of French Vegetables with Sulphate of Copper." This report gives a history of the practice in France and a discussion, and finally concludes:

We are of the opinion that the process of regreening is essentially fraudulent in its intention and commercial results; that regreening with sulphate of copper certainly does not make vegetables more wholesome—probably makes them less wholesome, and in some proportions always does so; that the public in purchasing preserved vegetables should call for preserved vegetables free from salts of copper; that the local authorities, as guardians of the public health, ought to come to no understanding as to the sale of vegetables containing sulphate of copper, but hold themselves free to act according to the circumstances of the case and the scientific evidence to be had from time to time.

This report was signed by the medical officer of health, James B. Russell; Peter Fyfe, sanitary inspector, and R. R. Tatlock and John Clarke, city analysts, and was dated September 17, 1890. Probably the market gardeners were not without influence in framing this report, as may be judged from the remark which also occurs in the report:

* * * the cultivator of the genuine fresh green vegetables is grossly prejudiced by the substitution for the produce of our market gardens of the last season's growth of foreign market gardens, colored so as to mislead the eye.

There is a good deal of truth in this complaint.

COPPER-GREENING IN THE UNITED STATES.

The subject of copper-greened vegetables has never attracted much attention in this country. In 1889 and again in 1891, the Massachusetts board of health caused analyses to be made of a large number of French preserved vegetables, and, finding copper, ordered cessation of their sale in the State. The reports are given on page 1159.

For a number of years the Brooklyn board of health has paid some attention to the presence of coppered foods on the Brooklyn markets, and has published a number of reports from chemists and others on the subject. In the annual report for 1887 there is a history of the coppering practice, written by Dr. W. H. Kent. In 1885 dealers were forbidden to sell pickles colored with copper (see page 1159). Shortly after the issuing of this prohibition a death occurred, alleged to be, according to the report of the board,[1] from the use of pickles so colored. The victim was Miss Maggie Martin, of 97 Adelphi street, Brooklyn.

The New Jersey dairy commissioner has also caused a number of analyses of imported canned vegetables to be made. Copper was discovered in many samples. (See page 1161.)

Packers in the United States use the copper process to some extent.

ANALYTICAL DATA.

Samples Bought.

All the samples whose analyses are recorded in this bulletin were bought at retail. Where it is not otherwise specified the samples were bought in Washington. A few were bought in towns in Florida, and a few in Schuyler, Nebr., the buying in both instances being done by employés of the Chemical Division of the Department of Agriculture. This was also the case with the samples purchased in this city. Full retail price was paid in each instance. No particular effort was made to procure either old or fresh samples, it being desired to get samples fairly representing the character of the canned vegetables on sale.

In the statements of the quantities of metallic contaminations found in these goods the amounts are invariably given as the number of milli-

---

[1] Annual report of Brooklyn Board of Health, 1885, 140.

grams of the metal in question in a kilo of the material as it came from the can. In some cases calculations are given of the quantities contained in a single can, and in this instance also the figures refer to the whole contents of the can, not to the vegetables alone.

PEAS.

There were 81 samples of canned and bottled peas examined. Of these 43 were packed in this country and the remaining 38 were labeled or sold as being of foreign packing. All these foreign samples were French, with one exception, No. 10719, which came from Italy. In the subjoined table are given the amounts of copper found in the French peas. The quantities of the metal given represent milligrams per kilo (parts per million) of the total contents of the bottle or can. This is the customary method of calculating, but in France another is sometimes used by which the copper is estimated only in the peas, the surrounding liquor being poured off before the analysis. This is done on the assumption that all copper is present in the peas and that none exists in the surrounding liquor.

Copper in French peas.

No.	Copper.	No.	Copper.	No.	Copper.	No.	Copper.
	mg.		mg.		mg.		mg.
10629[1][4]	0.0	10724	73.0	10879	66.2	10894	65.8
10661	157.7	10870[2]	0.0	10880	85.3	10895	55.5
10715[1]	79.2	10871	21.9	10885	99.2	10896	24.6
10716[1]	42.7	10872	61.9	10886	31.6	10897	41.9
10717	15.8	10873[2]	31.2	10887[4]	71.8	10903	17.1
10718	35.4	10874	39.0	10889	83.5	10904	65.8
10720	131.2	10875	77.8	10890[3]	61.4	10907	28.3
10721[3]	53.0	10876	34.8	10891[4]	127.4		
10722	28.5	10877	73.1	10892	78.6		
10723	28.0	10878	128.0	10893[4]	84.6		

[1] Colored with zinc.   [3] "Au naturel."
[2] Bore no label, but was sold as French peas.   [4] "A l'anglaise."

It will be seen from an inspection of this table that copper was present in every sample, with two exceptions, Nos. 10629 and 10870. The former of these samples was undoubtedly greened by the zinc process (see page 1051). The other when bought bore no label, but was sold as being "French peas." It may not have been packed abroad. In the other samples copper is present in amounts below Grimaux's proposed 18 mg limit in two cases.

In the American peas copper is not nearly so common, and in few instances does it occur in quantities sufficient to warrant the assumption that it entered otherwise than through the use of copper utensils. The occurrence of copper in these samples may be tabulated as follows:

Copper in American peas.

No.	Copper.	No.	Copper.	No.	Copper.	No.	Copper.
10625	51.0	10700	7.4	10711	20.9	10900	0.0
10626	4.8	10701	29.1	10712	25.0	10901	0.0
10627	0.0	10702	0.0	10713	0.0	10002	15.1
10628	2.1	10703	35.0	10714	3.4	10905	0.0
10059	7.0	10704	19.1	10681	74.1	10906	0.0
10694	4.6	10705	0.0	10882	44.0	10980	6.4
10695	4.6	10706	11.4	10883	9.0	10981	40.0
10696	1.8	10707	7.3	10884	0.0	10983	0.0
10697	1.6	10708	56.6	10888	12.8	10984	0.0
10698	0.0	10709	0.0	10808	16.5	10985	20.4
10699	6.5	10710	10.8	10890	0.0		

It will be seen from this table that 14 out of the 43 American peas were free from copper, 13 contained less than 10 mg and 5 more less than 18. In other words, there were altogether 32, which either contained no copper or contained it in an amount less than Grimaux's limit.

The comparison between the American and French peas in this regard may be tabulated as follows:

Copper in American and French peas.

	American.	French.		
	1	2	3	4
	Per cent.	*Per cent.*	*Per cent.*	*Per cent.*
Without copper	32.56	5.41	2.78	0.00
Containing copper	67.44	94.59	97.22	100.00
Containing copper, but less than 10 mg	30.23	0.00	0.00	0.00
Containing between 10 and 18 mg	11.63	5.45	5.56	5.74
Containing over 18 mg	25.58	89.19	91.66	94.20
Containing over 25 mg	16.28	83.78	85.28	88.57
Containing over 50 mg	6.98	56.76	58.33	60.00
Containing over 75 mg	0.00	29.73	30.56	31.43
Containing over 100 mg	0.00	10.81	11.11	11.43

Column 1 represents 43 samples; column 2 represents 37 samples, or all sold as French; column 3 represents 36 samples, or the number left after deducting the unlabeled sample; and column 4 represents 35 samples, or the number left after also deducting the zinc-greened sample.

The Italian sample of peas (No. 10719) contained copper to the amount of 14.6 mg. per kilo.

Fifteen out of the 81 pea samples contained salicylic acid. Five of these were French. The heavily coppered samples do not seem to receive additions of salicylic acid as a rule.

LABELS OF PEA SAMPLES. 1077

DESCRIPTION OF SAMPLES.

No. 10625. Hamburgh Champion of England. Hamburgh Canning Co., Hamburgh, N. Y.
This sample was bought from B. F. Bowen, Orlando, Fla., and cost 20 cents a can. The label read: "Hamburgh Champion of England peas. These peas are picked when in perfect condition and preserved by the 'Hamburgh process' in their original freshness, without artificial coloring. Packed at Hamburgh, Erie Co., N. Y. Packed by Hamburgh Canning Co., T. L. Bunting, Sec'y., Hamburgh, N. Y."

On opening the can there was a slight outflow of gas, and the interior appeared badly corroded. No preservative was found in the contents. These peas may be "without artificial coloring," but they contain copper in large quantity, 51 mg per kilo being found. This is equivalent to 29.8 mg per can. Assuming the correctness of the French statement, by which the amount of copper necessary to color a kilo of peas is said to be 10 mg, this sample contained something of an excess of the metal. Zinc was not found, but tin (35.6 mg) and lead (17.5 mg) were present. Assuming that all the lead existed as solder and hence was not in solution, there was still the amount of 18 mg of tin in solution.

No. 10626. Hamburgh petits pois verts, extra fins. Hamburgh Canning Co., Hamburgh, N. Y. This sample was bought from B. F. Bowen, Orlando, Fla., and cost 30 cents. The label read: "Hamburgh petits pois verts, extra fins, Hamburgh Canning Company, Hamburgh, Erie County, N. Y. Hamburgh process. First pickings of the choicest variety of French sweet peas and pronounced superior in delicacy of flavor to the imported article."

The can appeared corroded on the interior. No preservative was found. In this sample the packers were not so liberal with copper as in No. 10625, since but 4.8 mg per kilo were found. There was no zinc, but there were 3.1 mg of lead and 10.2 mg of tin, showing that a little tin was certainly in solution.

No. 10627. Jumbo brand early June peas. Miller Brothers & Co., Baltimore. This sample was also bought from B. F. Bowen, Orlando, Fla., and cost 20 cents a can. The label read: "Early June peas, Jumbo brand. Packed by Miller Brothers & Co., at Baltimore, Md. All goods bearing our name are of first quality, carefully selected and packed for the finest family trade."

The label was adorned with a picture of a negro presenting an elephant an oyster. The can was corroded. Salicylic acid was found, but there was no copper, lead, or zinc. There were 39.4 mg of tin per kilo, which must all have been in solution.

No. 10628. Sifted early June peas. F. H. Leggett & Co., New York. This sample was bought from B. F. Bowen, Orlando, Fla., and cost 25 cents. The label read: "Sifted early June peas. Francis H. Leggett & Co., New York. Packed at Baltimore, Baltimore Co., Md."

On opening the can there was a slight outflow of gas. The can was corroded. There were no preservatives found. Copper was present in small quantity, 2.1 mg per kilo being found. Its presence was probably accidental. There was no zinc or tin, and but a trace of lead.

No. 10629. Petits pois surfins, à l'anglais. Vve. Garres, jne. & Fils, Bordeaux. This sample was bought of B. F. Bowen, Orlando, Fla., and cost 30 cents. The label, which was printed on the can, read: "Petits pois surfins, à l'anglais. Vve. Garres, jne. & Fils, Bordeaux, France."

No preservatives were found. There was no copper in this sample, but zinc was found to the amount of 85.5 mg per kilo, or 33.7 mg per can. It appears probable that this was one of the zinc-greened samples, the manufacture of which is mentioned by Gautier (see p. 1051). The use of zinc in the greening process is even more objectionable than that of copper.

No. 10659. Green peas, little fellows. Numsen & Sons, Baltimore. This can was bought in Kissimmee, Fla., and cost 25 cents. The label read: "Green peas, 'little fellows.' These peas do not require any cooking. Wm. Numsen & Sons, Baltimore, Md., established 1847. Clipper brand trade-mark registered January 14, 1879. First quality. Packed at Baltimore, Md." The label bore the picture of a ship at sea.

On opening the can there was a slight outflow of gas. The interior was slightly corroded. Salicylic acid was found to be present, and copper was also found, there being 7 mg per kilo. No zinc was detected, but lead and tin were present. Possibly they were in the form of solder.

No. 10661. Petits pois extra fins. Talbot Frères. Bordeaux. This can was bought in Kissimmee, Fla., and cost 30 cents. The label which was printed on the can read: "Petits pois extra fins verts. Talbot Frères, Bordeaux, France. Fabrique de conserves alimentaires."

No preservatives could be detected. The enormous amount of 157.7 mg of copper per kilo (64 mg per can) was found. Comment on this is unnecessary. Tin was also present in this sample. Copper was found in Talbot Frères' peas by the Massachusetts State board of health in 1891 and in their beans in 1889.

No. 10694. Marrowfat peas. Parson Bros., Aberdeen, Md. One can bearing this brand was bought from A. A. Winfield, 215 Thirteen-and-a-half street SW., and cost 10 cents. Another was bought from Frank Hume, 454 Pennsylvania avenue, and cost 15 cents. The label read: "Marrowfat peas, first quality. Packed by Parson Bros., Aberdeen, Maryland. Parson Bros. brand. 1889."

On opening the can there was a slight outflow of gas. The can was corroded. Salicylic acid was found, and copper to the amount of 4.6 mg per kilo was also present. Of zinc there were 12.5 mg per kilo. There was also a trace of lead.

No. 10695. AA brand early June peas. J. F. Lowekamp, Jessups, Md. One can of this sample was bought from A. A. Winfield, 215 Thirteen-and-a-half street SW., and cost 10 cents. Another was bought from C. E. Nelson, corner Seventh and E streets SE., at a cost of 13 cents. The label read: "Early June peas, the Queen Anne, AA brand. Best quality. Packed at Jessups, Anne Arundel Co., Md., by J. F. Lowekamp."

On opening the can there was a slight outflow of gas. The can appeared to be corroded. No preservative could be detected. Copper to the amount of 4.6 mg per kilo was found, but there was no zinc or lead. Of tin 27 mg per kilo were found. The presence of the copper was probably accidental.

No. 10696. Cumberland brand sifted early June peas. J. T. Cox, Bridgeton, N. J. One can of this brand was bought from G. C. Burchard, 354 Pennsylvania avenue, and cost 13 cents. Another came from Browning & Middleton, 610 Pennsylvania avenue, and cost 10 cents. The label read: "Early June peas, sifted; Cumberland brand; first quality. Packed by John T. Cox, Bridgeton, Cumberland County, N. J."

No preservative was found. There was a small amount of copper, 1.8 mg being present, but no zinc. Lead and tin were found in large quantities, but may have been present as solder.

No. 10697. Early June peas. Chas. Brewington & Co., Baltimore. This sample was bought from G. C. Burchard, 354 Pennsylvania avenue, and cost 13 cents. The label read: "Early June peas, first quality. Packed by Chas. Brewington & Co., at Baltimore, Md." Trade mark, the monogram C. B., with the word "brand" crossing it.

The can was slightly corroded. No preservative was found, but there was a little copper (1.6 mg). No zinc was detected. The copper was very likely introduced accidentally, as the quantity was small.

No. 10698. Chester River brand early June peas. Martin Wagner Co., Baltimore. One can was bought from G. C. Burchard, 354 Pennsylvania avenue, and cost 13 cents. Another came from H. I. Meador, corner of Eighth and I streets SW., and cost 15 cents. The label read: "Early June peas, Chester River brand. Packed by Martin Wagner Co., at 2315, 2317 Boston street, Baltimore City, Md."

On opening there was a slight outflow of gas. The interior of the can was slightly corroded. No preservative was found. Copper was found to the amount of 9 mg per kilo. Zinc was present in somewhat larger quantity, 10.9 mg. There was a trace of lead.

No. 10699. Silver brand early June peas. B. F. Shriver & Co., Union Mills, Md. One can was bought from A. A. Winfield, 215 Thirteen-and-a-half street SW., and cost 13 cents. Another came from Jackson & Co., 626 Pennsylvania avenue, and cost 15 cents. The label read: "Early June peas, silver brand. B. F. Shriver & Co., Union Mills, Carroll Co., Md."

No preservative could be detected. There were 6.5 mg of copper per kilo in this sample, but no zinc. Lead was found to be present to the extent of 4.2 mg, and tin to that of 21.6 mg, showing that there was some tin in solution.

No. 10700. Franklin brand marrowfat peas. Franklinville Canning Co., Franklinville, N. Y. One can was bought from J. T. Earnshaw, corner of Eighth and G streets SE., and cost 10 cents. Another came from J. B. Bryan & Bro., 608 Pennsylvania avenue NW., and cost 12 cents. The label read: "Franklin brand marrowfat peas, first quality; Franklin peas. Packed by the Franklinville Canning Co., at Franklinville, Cattaraugus Co., N. Y." The label bears a picture of Franklin.

The can was slightly corroded. No preservatives could be detected. Some copper was present, 7.4 mg being found. There was no zinc, but 5.1 mg of lead and 19.9 of tin were found. Some tin was evidently in solution.

1080 FOODS AND FOOD ADULTERANTS.

No. 10701. Centennial brand early June peas. A. B. Roe, Greensborough, Md. One can was bought from Estler Bros. & Co., corner of C and Thirteenth streets SW., and cost 13 cents. One was also bought of N. H. Shea, 632 Pennsylvania avenue, and cost 15 cents. The label read: "Early June peas, Centennial brand. A. B. Roe. Choice quality, packed for select trade. Fresh from the farm at Greensborough, Caroline Co., Md. The contents of this can were packed with great care by A. B. Roe, at Greensborough, Caroline Co., Md." The label bore a picture inscribed "Independence Hall, Phila., 1776.'

On opening the can there was a slight outflow of gas. The can was slightly corroded. No preservative was found. Copper was present, 29.1 mg. being found. There was no zinc. Lead was present in small amount, and tin in large quantity.

No. 10702. Blue Ridge brand early sweet peas. B. F. Shriver & Co., Union Mills, Md. This sample was bought from Jackson & Co., 626 Pennsylvania avenue, and cost 10 cents a can. The label read: "Early sweet peas, Blue Ridge brand. B. F. Shriver & Co., Union Mills, Carroll Co., Md."

This sample, like others of Shriver & Co.'s goods, contained salicylic acid. There was no copper or zinc. Lead and tin were present in some quantity, but probably existed as solder.

No. 10703. Extra small peas. B. F. Shriver & Co., Union Mills, Md. This can was bought of N. H. Shea, 632 Pennsylvania avenue NW., and cost 18 cents. The label read: "Extra small peas. Packed by B. F. Shriver & Co., Union Mills, Carroll Co., Md."

The can appeared to be corroded. Salicylic acid was detected. Copper was present in these peas, 35 mg per kilo, or 21.4 per can being found. Lead and tin were both present, but probably as solder. There was no zinc.

No. 10704. Marrowfat peas. Fait & Winebrenner, Baltimore. This sample was bought of N. H. Shea, 632 Pennsylvania avenue NW., and cost 10 cents. The label read: "Marrowfat peas, Nunley, Hynes & Co., Baltimore, Md. Packed by Fait & Winebrenner, at Baltimore, Baltimore Co., Md." The word "soaked" is given in an almost illegible combination of letters.

On opening the can there was a slight outflow of gas. The can was corroded. No salicylic acid or other preservative was found. Copper to the amount of 19.1 mg per kilo was present. There were 38.1 mg of zinc per kilo. There was also a trace of lead.

No. 10705. Marrowfat peas. T. W. Clark & Son, Glenville, Md. This sample was bought of N. H. Shea, 632 Pennsylvania avenue, and cost 12 cents. The label read: "Marrowfat peas; first quality. Thos. W. Clark & Son, Glenville, Md. Packed by Fait & Winebrenner, at Baltimore, Baltimore Co., Md."

The can was somewhat corroded. No preservative could be detected. No copper was present, nor was zinc or lead. Tin was found to the extent of 50 mg per kilo.

No. 10706. Small May peas, Clipper brand. Wm. Numsen & Sons, Baltimore. This sample was bought from Browning & Middleton, 610 Pennsylvania avenue NW., at a cost of 15 cents a can. The label read: "Small May peas; first quality; Clipper brand, (established 1847). Packed at Baltimore, Md., by Wm. Numsen & Sons. Trade-mark registered Jan. 14, 1879. These peas do not require any cooking." The label also bears a picture of a ship at sea.

On opening the can there was a slight outflow of gas. No preservative was found. Copper was present, 11.4 mg being found. Zinc was absent; lead and tin were present, but probably as solder, the proportions being nearly one to one.

No. 10707. Early June peas. T. J. Myer & Co., Baltimore. This sample was bought from J. J. Daly, 1367 C street SW., and cost 15 cents. The label read: "Early June peas, first quality. Packed by T. J. Myer & Co., Baltimore, Baltimore Co., Md." The label also bore as a trade-mark the arms and motto (crescite et multiplicamini) of the State of Maryland.

On opening the can there was a slight outflow of gas. The can was somewhat corroded. No preservative could be detected. Copper, 7.3 mg per kilo. Zinc was present, 31.2 mg being found. Lead was present in very small quantity.

No. 10708. Early June peas. Gibbs Preserving Co., Baltimore. This sample was bought from S. S. Tucker, corner Thirteenth and C streets SW., and cost 13 cents. The label read: "Early June peas. Packed at Baltimore, Baltimore Co., Maryland, by Gibbs Preserving Co. First quality." The trade-mark was a bull's head.

On opening the can there was a slight outflow of gas. The can showed some corrosion. Salicylic acid was found to be present. There was a large amount of copper, 56.6 mg per kilo being found. This is a larger quantity than some of the French packers use. There was no zinc. Lead and tin were present in nearly equal amounts, and therefore probably existed as solder.

No. 10709. Van Camp's sifted peas. Van Camp Packing Co., Indianapolis. This sample was bought from Browning & Middleton, 610 Pennsylvania avenue, and cost 10 cents. It was labeled: "Van Camp's sifted peas." On a picture of a stamped envelope postmarked "New York" was the address "Van Camp Packing Co., Indianapolis, Ind."

On opening the can there was a slight outflow of gas. The can was corroded and the contents emitted an unpleasant odor. No preservative was present, nor was copper found. There were 32.4 mg of zinc per kilo or 20.4 mg per can. Lead and tin were present, but probably as solder.

No. 10710. Gold leaf brand sweet early June peas. Western New York Preserving & M'f'g. Co., Springville, N. Y. This sample was bought from N. H. Shea, 632 Pennsylvania avenue NW., and cost 15 cents. The label read: "Sweet early June peas, gold leaf brand. Packed at Springville, Erie Co., N. Y., by the Western New York Preserving & M'f'g Co." The label is also embellished with a map of Erie County and vicinity, and a picture of a girl's head.

No preservatives could be found. There were 10.8 mg of copper per kilo, and, in addition, zinc to the enormous amount of 380 mg. Lead in the amount of 13.9 and 32.6 mg of tin were also present. Some of the tin was in solution, even if an amount equal to the lead be reckoned as being in the form of solder.

No. 10711. Early June peas. W. S. Whiteford, Delta, Pa. This sample came from N. H. Shea, 632 Pennsylvania avenue, and cost 12 cents. The label read: "Early June peas. W. S. Whiteford, Delta, Pa."

The can was somewhat corroded. The contents did not appear to have received an addition of a preservative. As to metallic contaminations this sample did not appear so favorably, there being present 20.9 mg of copper and 16 mg of zinc per kilo. Lead and tin were also present, but being in nearly equal quantities, may have existed in the shape of solder.

No. 10712. Marrowfat peas. G. W. Hunt & Co., Baltimore. This sample was bought from J. T. Earnshaw, 534 Eighth street SE., and cost 10 cents. It was labeled: "Marrowfat peas, G. W. Hunt & Co., Baltimore, Md." The label bore the monogram of the packer.

The can was slightly corroded. No preservative could be found. Copper was present to the amount of 25.0 mg per kilo and zinc to that of 6.7 mg. Of the lead there were 11.5 mg, and also 24.2 of tin, showing that some of the latter metal was in solution.

No. 10713. Harpoon brand early June peas. J. Ludington & Co., Baltimore. This sample was also bought of J. T. Earnshaw, 534 Eighth street SE., and cost 15 cents. It was labeled "Early June peas, harpoon brand; first quality. J. Ludington & Co., Baltimore, Md. All goods under this brand are of our own packing and warranted to give entire satisfaction."

On opening the can there was a slight outflow of gas, and it was found to be slightly corroded. No preservative could be found, nor could copper or zinc. Lead and tin were present in nearly equal amounts, and therefore possibly in the shape of solder.

No. 10714. Harpoon brand extra sifted peas. J. Ludington & Co., Baltimore. This sample was bought from J. T. Earnshaw, 534 Eighth street SE., and cost 18 cents. The label read: "Extra sifted peas, harpoon brand; first quality. J. Ludington & Co., Baltimore, Md. All goods under this brand are of our own packing and are warranted to give entire satisfaction."

The can was slightly corroded. The peas were yellow, though a little copper was present, the amount being 3.4 mg per kilo. This small amount may have gained access through accident during the canning. There was no zinc found, but lead and tin were both present. Salicylic acid was also present.

No. 10715. Petits pois fins, au naturel. A. Nadal, Bordeaux. This sample was bought from Frank Hume, 454 Pennsylvania avenue NW., and cost 30 cents a can. The label, printed in blue and gold, read: "Petits pois fins au naturel. Amédée Nadal, successeur de Congouille, Bordeaux. Usine à Eymet (Dordogne)." The remark: "We guarantee our packing to be of extra superior quality, A No. 1," is given in both English and French. There is also a list of medals of honor, etc., gained at various expositions by the packer.

The can was not corroded. Salicylic acid was present. The peas were of a bright green color, evidently due to the copper which was present in abundance, there being 79.2 mg per kilo or 34.4 per can. This packer is usually lavish with the metal. There was no zinc, but a little lead and tin were found.

No. 10716. Petits pois moyens, au naturel. A. Nadal, Bordeaux. This sample came from Frank Hume, 454 Pennsylvania avenue NW., and cost 25 cents. The label (in green and gold) read: "Petits pois moyens au naturel. Amédée Nadal, successeur de Cougouille, Bordeaux." The rest of the label is a duplicate of that of the preceding sample.

This sample was from the same packer as the preceding one. It differed from it in containing no preservative and in being less rich in copper, there being 42.7 mg per kilo, or 44.1 mg per can. Any deficiency in this respect, however, was more than compensated for by the zinc, of which 101.0 mg per kilo were found. There was also some lead and a large amount of tin.

No. 10717. Petits pois. Jules Dupont, Paris. One can was bought from Estler Bros. & Co., corner C and Thirteenth streets SW.; one from J. B. Bryan & Bro., 608 Pennsylvania avenue NW.; one from Beall & Baker, 486 Pennsylvania avenue NW., and one from Frank Hume, 454 Pennsylvania avenue NW. The prices were respectively: 25, 25, 10, and 15 cents. The label read: "Petits pois. Jules Dupont, Paris." The label was of metal and soldered to the can.

The can was slightly corroded. No preservative was present. The peas were green. Copper to the extent of 15.8 mg per kilo was present, but there was no zinc. Lead and tin, also present, being in nearly equal quantities, possibly existed as solder.

No. 10718. Petits pois fins. G. Triat & Co., Bordeaux. The sample was bought from G. C. Burchard, 354 Pennsylvania avenue, NW. It cost 20 cents a can. The label read: "Excelsior petits pois fins. Carefully packed for first-class trade. Gabriel Triat & Co., Bordeaux."

The can was slightly corroded. No preservative was found. The peas were green, the color being due to copper, of which 35.4 mg per kilo were found. There was no zinc. Tin and lead were both present, but in nearly equal quantity, and therefore possibly existed in the form of solder.

No. 10719. Petits pois fins. F. Cirio, Turin. This sample was bought from Frank Hume, 454 Pennsylvania avenue, at a cost of 15 cents. The label was: "Petits pois fins. Francesco Cirio, Turin. Diplôme d'honneur et médaille d'or, Paris. Conserves alimentaires. Fournisseur des cours impér. et roy."

The can was slightly corroded. No preservative was found. The peas were coppered with 14.6 mg per kilo and were green. No zinc was found, but tin and lead, possibly as solder, were found in some amount.

No. 10720. Petits pois moyens. C. Couteaux, Paris. This sample was bought from Frank Hume, 454 Pennsylvania avenue, and cost 15 cents. The label was: "Petits pois moyens. C. Couteaux, Paris."

The can was slightly corroded. The peas were green and contained no preservative. There was no less an amount of copper than 131.2 mg per kilo present. This is 55.1 mg per can. Zinc was also found in slight amount, there being 3.1 mg per kilo. Lead and tin were present, possibly as solder.

No. 10721. Petits pois extra fins, au naturel. A. Nadal, Bordeaux. This sample was bought from J. B. Bryan & Bro., 608 Pennsylvania avenue, and cost 30 cents. It was labeled: "Petits pois extra fins, au naturel. Amédée Nadal, successeur de E. Congouille, Bordeaux. Usines Eymet, Dordogne." The rest of the label is like that of No. 10715.

The can was slightly corroded. There was no preservative found. The peas were very green, and contained 53.0 mg of copper per kilo. The packer evidently does not regard the " au naturel " recommendation of M. Grimaux (page 1064) as a thing to be observed. There was no zinc and but little lead, though considerable tin.

No. 10722. Petits pois fins. Dandicolle & Gaudin, Bordeaux. This sample was bought from J. B. Bryan & Bro., 608 Pennsylvania avenue NW., at a price of 25 cents. The label was: "Petits pois fins. Dandicolle & Gaudin, Bordeaux."

The can was corroded. No preservative was found. The peas were green and contained copper to the extent of 28.5 mg per kilo. There was no zinc. Lead and tin were present, the latter in the larger quantity.

No. 10723. Petits pois extra fins. Gobelin Fils & Cie, Bordeaux. This sample was bought from Beall & Baker, 486 Pennsylvania avenue, and cost 25 cents. The label was: "Petits pois extra fins. Gobelin Fils & Cie, Bordeaux, France. Conserves alimentaires de qualité extra."

The can bore a complicated design in the nature of a trade mark, representing a banner with the monogram "G. F." and an eagle perched above. The can was corroded. No salicylic acid or other preservative was found. Copper was present to the extent of 28.0 mg per kilo.

No. 10724. Petits pois. Tisserand et Fils, Paris. This sample was purchased of Jackson & Co., 626 Pennsylvania avenue NW., at a cost of 20 cents. The label was "Petits pois. Tisserand et Fils, Paris."

The can was corroded. No preservative was found. The bright green color of the peas was due to the presence of a high amount of copper, there being present 73.0 mg per kilo or 35.6 mg per can. Lead and tin were present, but there was no zinc.

No. 10870. No label. This sample was bought from William T. Davis, corner of Fifteenth and P streets NW., at a cost of 15 cents. It bore no label whatever, but was sold as containing French peas.

The can showed a slight corrosion and the contents a slight green color. No copper or zinc was present, however. There was no preservative.

No. 10871. Petits pois fins. Vve. Aubin-Salles, Nantes. This sample was bought of E. E. Berry, stand 1, Riggs Market, and cost 25 cents. The label was: "Petits pois fins. Vve. Aubin-Salles, Ville-en-bois, Nantes. Exploitation à proximité de mes usines de deux vastes fermes pour la culture et la garantie de fraicheur des légumes employés pour mes conserves de petit pois, haricots verts, etc., etc. Produits garantics."

The can was but slightly corroded. Salicylic acid was present. The peas, which were a faint green, showed 21.9 mg of copper per kilo. No zinc or lead was found.

LABELS OF PEA SAMPLES. 1085

No. 10872. *Petits pois fins.* *Couteau, Paris.* This sample was bought of Hagan & Gilkison, stand 60, Riggs market, and cost 18 cents. The label was: "Petits pois fins. Couteau, Paris."

The can was slightly corroded. No preservative was found. The peas were green in color and contained 61.9 mg of copper and 13.0 mg of zinc per kilo. Tin and lead were also present, the former in much larger quantity than the latter.

No. 10873. *Petits pois extra fins au naturel.* *Amieux Frères, Nantes and Paris.* This sample was also bought from Hagan & Gilkison, stand 60, Riggs market, at a cost of 18 cents. The label bore a complicated design and the punning inscription: "Petits pois extra fins gastronomes au naturel. Dans le choix des conserves la marque seule doit guider l'acheteur, qui ne peut en apprecier la qualité qu' au moment de les consomer. Nantes et Paris. Notre devise est comme notre nom: 'Toujours à mieux.' Amieux Frères. Quatre medailles d'or, etc."

The can was slightly corroded. There was no preservative found. The peas were yellow, but contained 31.2 mg of copper and 58.9 mg of zinc per kilo. The products of these packers are not quite so praiseworthy as their label represents. Tin and lead were also present.

No. 10874. *Petits pois moyens.* *V. Leopold, Bordeaux.* This sample was bought of F. E. Altemus, 1410 P street NW., and cost 15 cents. Label: "Petits pois moyens. Victor Leopold, Bordeaux, France."

The can was slightly corroded. No preservative was found. The peas, which were green, contained 39.0 mg of copper per kilo. There was no zinc, but lead was present.

No. 10875. *Petits pois fins.* *Gobelin Fils & Cie, Bordeaux.* This sample likewise came from F. E. Altemus, 1410 P street NW., and cost 25 cents. Label: "Petits pois fins. Conserves alimentaires de qualité extra. Gobelin Fils & Cie, Bordeaux, France." The monogram of the firm forms a trade-mark.

The can showed slight corrosion. The contents were green in color, and contained 77.8 mg of copper and 97.4 mg of zinc per kilo. The cause of the color of the peas is not hard to find. Lead was present. No salicylic acid was detected.

No. 10876. *Petits pois moyens.* *Robinet & Cie, Nantes.* This sample was bought from Birch & Co., 1414 Fourteenth street NW., and cost 12 cents. It was labeled: "Petits pois moyens. Robinet & Cie, Nantes, France."

The can was slightly corroded. No preservative was found. The peas were green and contained 34.8 mg of copper per kilo. There was no zinc.

No. 10877. *Petits pois extra fins.* *L. A. Price, Bordeaux.* This sample was bought from Birch & Co., 1414 Fourteenth street NW., and cost 35 cents a bottle. The label read: "Petits pois extra fins. L. A. Price, Bordeaux, France. Conserves de qualité extra supérieure. Produit medaillés."

The can was only slightly corroded. No preservative was detected. The peas were only faintly green, but nevertheless contained 73.1 mg of copper per kilo. There was no zinc, tin, or lead.

No. 10878. Petits pois. Le Lagadec, Lorient, France. This sample was procured of M. F. Crown, 1532 Fourteenth street NW., and cost 20 cents. It was labelled: "Petits pois. Le Lagadec, Lorient, France."

The can was corroded. No preservative was found. The peas appeared to be full grown and were of a dark green color, due to the presence of an inordinate amount of copper, 128.0 mg per kilo, or 55.9 mg per can being found. There was no zinc. In view of the amount of copper present it was probably not thought necessary to add it.

No. 10879. Petits pois extra fins. E. Du Raix, Bordeaux. This sample was bought of John P. Love, 1534 Fourteenth street NW., and cost 35 cents. It was put up in a glass bottle with a lead top, and with nothing to intervene between the peas and the lead. It was labeled: "Petits pois extra fins. Eugène Du Raix, Bordeaux. Conserves extra."

The peas were very small and green. No preservative was found. There were 66.2 mg of copper per kilo, or 27.9 per bottle. There was no zinc or tin. Lead to the enormous amount of 35.2 mg per kilo, or 14.8 mg per bottle, was present. It was, of course, derived from the lead top. The use of this style of package displays the most flagrant disregard of the laws of hygiene. Goods thus preserved should not be allowed to enter our ports. It seems incomprehensible that the health authorities of Bordeaux should allow the two firms, Du Raix and Dandicolle & Gaudin, to bottle food in this way.

No. 10880. Petits pois extra fins. Henri Lambert & Cie, Bordeaux. This sample was also bought of John P. Love, 1534 Fourteenth street NW., and cost 35 cents. It was labeled: "Petits pois extra fins. Henri Lambert & Cie, Bordeaux, France."

No preservative was found. These were very small green peas and contained much copper, 85.3 mg per kilo being found. This is equal to 35.7 mg per can. There was no zinc. Tin was present in some quantity, but there was no lead.

No. 10881. Peninsular brand June peas. W. C. Satterfield, Greensborough, Md. This sample was bought of John P. Love, 1534 Fourteenth street NW., for 15 cents. It was labeled: "Standard June peas, Peninsular brand. Packed by W. C. Satterfield, Greensborough, Caroline Co., Md." A scroll map of Greensborough and vicinity forms a part of the label.

No preservative was found. These peas were large-sized and yellow in color, apparently having been mature at the time of packing. Notwithstanding their color, both copper and zinc were present and in large quantity. Of the former there were 74.1 mg per kilo and of the latter 163.0 mg. There was a trace of lead. Comment on this sample is superfluous.

No. 10882. Sifted early June peas. Jas. Wallace & Son, Cambridge, Md. This sample was bought of John W. Hardell, 1428 Ninth street, at the rate of 12½ cents per can. The label was: "Abbsco brand sifted Early June peas. Packed by Jas. Wallace & Son at Cambridge, Dorchester Co., Md. First quality."

These peas were medium in size, but appeared to be mature. Notwithstanding the fact that they were yellow in color, copper to the extent of 44.0 mg per kilo and zinc to that of 15.3 mg were present. Traces of tin and lead were also found.

No. 10883. Early June peas. J. Ludington & Co., Baltimore. This sample was bought from J. H. Hungerford, 1334 Ninth street NW. Price 18 cents. Label: "Early June peas, J. Ludington & Co., Baltimore."

These peas were medium-sized and yellow colored. They contained no preservative, copper, or zinc, and but a trace of lead.

No. 10884. Sifted early June peas. Chas. Laing & Co., Baltimore. This sample was bought from J. H. Hungerford, 1334 Ninth street NW., at a cost of 20 cents. Label: "Sifted early June peas, first quality. Charles Laing & Co., Baltimore, Baltimore Co., Md. All goods bearing our name guaranteed to be of superior quality."

The peas contained in this can were small and green, but contained neither copper nor zinc. Salicylic acid was found.

No. 10885. Petits pois extra fins. Dandicolle et Gaudin, Bordeaux. This sample came from J. H. Hungerford, 1334 Ninth street, and cost 40 cents. It was labeled: "Petits pois extra fins. Dandicolle et Gaudin, 'Limited,' Bordeaux, France."

This sample was contained in a bottle with a lead top, one of the style affected by this firm. The peas were small and green, and showed the enormous amount of 99.2 mg of copper per kilo. There was no zinc and but a trace of lead. From the nature of the package this absence of lead was peculiar. A large amount of salicylic acid was found.

No. 10886. Petits pois extra fins. M. Ader & Cie., Paris. Bought of Robert White, jr., 900 Ninth street NW. Price 25 cents. Label: "Petits pois extra fins. M. Ader & Cie., Paris. Conserves alimentaires, perfectionnées garanties."

These peas were of medium size. Salicylic acid was present in large quantity. Copper was also present, 31.6 mg per kilo being found.

No. 10887. Petits pois fins à l'anglaise. Vve. Garres, jne. & Fils, Bordeaux. This sample was bought from Robert White, jr., 900 Ninth street NW., and cost 25 cents. It was labeled: "Petits pois fins à l'anglaise. Vve Garres, jne. & Fils, Bordeaux, France." The label also gives full directions for cooking.

No preservative was found. There were 71.8 mg of copper per kilo in this sample, but no zinc, so that it was materially different from another sample (No. 10629) put up by this packer and of nearly the same label, in which there was an enormous amount of zinc, but no copper. The difference in the labels consisted in the presence of "sur" before the word "fins" in sample No. 10629.

No. 10888. Wholesome brand early June peas. A. W. Sisk, Preston, Md. This sample was purchased from Robert White, jr., 900 Ninth street NW., and cost 10 cents. The label read: "Early June peas, wholesome brand. Packed by A. W. Sisk, at Preston, Caroline Co., Md. Established 1887. Extra quality." The label also bears the picture and signature of the packer.

The contents of the can in this sample were covered with a white mucilaginous substance, and the peas themselves were yellow. The packer may regard copper as being necessary for a "wholesome brand," for there were 12.8 mg per kilo of this metal present. No zinc was found, nor could a preservative be detected. Lead and tin, possibly in the shape of solder, were present.

No. 10889. Pois moyens. Baron, Père et Fils, Paris. Bought of J. F. Russell, 730 Ninth street NW., at a cost of 16 cents. The label read: "Pois moyens, Baron, Père & Fils, Paris."

No preservative was found. The peas were medium in size and green in color. They contained 83.5 mg of copper per kilo and 29.0 mg of zinc. There was some tin in solution.

No. 10890. Petits pois moyens au naturel. J. Nouvialle & Cie., Bordeaux. Bought of J. F. Russell, 730 Ninth street, at a price of 25 cents. Label: "Petits pois moyens au naturel. J. Nouvialle & Cie., Bordeaux." Nouvialle & Cie. also claim the possession of any number of medals, etc.

No preservative was present. The peas, which were of medium size and green in color, contained 61.4 mg of copper per kilo. There was no zinc, the packer probably being of the opinion that it was not necessary in the presence of sufficient copper.

No. 10891. Petits pois extra fins à l'anglaise. Amieux Frères, Nantes. This sample was bought of John Keyworth, 318 Ninth street, and cost 30 cents. It was labeled: "Petits pois extra fins à l'anglaise, Amieux Frères, Nantes, France. Notre devise est comme notre nom: Toujours à mieux."

An enormous amount of salicylic acid was found in this sample. The peas were small and green and contained 127.4 mg of copper per kilo. This firm, in its anxiety for improvement, rather goes to an extreme in the use of copper. Zinc was not present.

No. 10892. Petits pois fins. Geo. Cadeau & Cie. This sample was bought from John Keyworth, 318 Ninth street, and cost 30 cents. The label was: "Petits pois fins. Geo. Cadeau & Cie. Produits perfectionnés et garantis 1re qualité. All our goods are warranted to be prepared in their natural form without any adulteration what-so-ever."

No preservative was found. The peas were small and covered with a thick juice. Notwithstanding the guaranty of the packers, coppering had been practiced. The metal was present to the extent of 78.6 mg per kilo. There was no zinc, but lead was found.

No. 10893. Pois à l'anglaise. S. Nicolas & Cie., Bordeaux. The sample was bought from Elphonzo Youngs Co., 428 Ninth street, and cost 16 cents. The label read: "Pois à l'anglaise. S. Nicolas & Cie., Bordeaux, France."

These peas were of a large size and green color. They did not contain preservatives, but were not lacking in copper. Of this metal 84.6 mg per kilo, or 35.6 mg per can, were found. There was no zinc and but a trace of lead.

No. 10894. Petits pois extra fins. Guillaumez, Nancy. This sample was bought from Elphonzo Youngs Co., 428 Ninth street, and cost 25 cents. The label read: "Petits pois extra fins. Crown Imperial, Lunéville. Guillaumez, Nancy. Fabrication perfectionnée. Usine à vapeur. Produits du pays, fruits, légumes, etc."

This can contained small green peas. No preservative was found. Coloring had been done with copper, the quantity of 65.8 mg of the metal per kilo, or 29 mg per can, being present. There was no zinc, but lead and tin were present, the latter in the greater quantity.

No. 10895. Petits pois. Guillaumez, Nancy. This sample was bought from Elphonzo Youngs Co., 428 Ninth street, at a price of 20 cents. The label read: "Petits pois, fabrication perfectionnée." The rest of the label was like that of the preceding sample.

Salicylic acid was present. The peas, which were very small and green, contained per kilo 55.5 mg of copper. Other metallic contaminations were absent.

No. 10896. Petits pois extra fins. J. Fiton, Aîné & Cie., Bordeaux. This sample was bought from J. F. Russell, 730 Ninth street, and cost 16 cents. The label read: "Petits pois extra fins. Conserves alimentaires perfectionnées et garanties. J. Fiton, Aîné & Cie.," with the usual list of prizes and medals.

No preservative was found. The peas were green and very small. Copper was present to the extent of 24.6 mg per kilo. There was no lead or zinc, but tin existed in large quantity.

No. 10897. Petits pois extra fins. J. Fiton, Aîné & Co., Bordeaux. This sample came from Elphonzo Youngs Co., 428 Ninth street, and cost 35 cents. The label read: "Petits pois extra fins. J. Fiton, Aîné & Co., Bordeaux, 183 and 185 Rue Fondaudège. Fabrication extra supérieure." This sample differed from the preceding in being put up in a bottle instead of a can.

Salicylic acid was found to be present in large quantity. Copper greening had been practiced with these peas. The quantity of the metal found was 41.9 mg per kilo. Zinc was absent but lead was present in large amount, there being 29.2 mg per kilo or 12.5 mg per bottle.

No. 10898. Sweet blossom peas. Githens, Rexsamer & Co., Philadelphia. This sample was purchased of Elphonzo Youngs Co., 428 Ninth street, and cost 25 cents. The label read: "Sweet blossom peas; sweet and tender. Put up by Githens, Rexsamer & Co., Philadelphia, Pa."

No preservative was found. The peas were medium-sized and yellow. In spite of their color copper was present to the amount 16.5 mg per kilo. There was no lead or zinc. Tin existed in solution to the extent of 39.8 mg per kilo.

No. 10899. Royal favorite peas. Githens, Rexsamer & Co., Philadelphia. This sample was bought from Elphonzo Youngs Co., 428 Ninth street, and cost 30 cents. The label read: "Royal favorite peas. Delicately sweet, tender, and small. Packed by a new process, giving the royal favorite brand of peas a delicacy and sweetness unrivaled. Githens, Rexsamer & Co., sole agents, Philadelphia, Pa."

On opening the can there was an outflow of gas. The peas were small and yellow. No preservative was found, but there was a trace of copper. Zinc was not found. Tin and lead, possibly in the shape of solder, were present.

No. 10900. Extra fine pansy blossom peas. Keagle & Guider, Baltimore. This sample came from Elphonzo Youngs Co., 428 Ninth street, NW., and cost 12¼ cents a can. The label read: "Extra fine pausy blossom peas; marrowfats; Maryland Jockey Club brand. First quality. Packed at Baltimore. Keagle & Guider, packers, Baltimore, Md."

There was no preservative. The peas were yellow and evidently mature when packed. Lead, copper, and zinc were absent. Tin, however, was present.

No. 10901. Early June peas. H. P. Cannon, Bridgeville, Del. This sample was bought from C. I. Kellogg, 602 Ninth street, NW., and cost 15 cents. The label read: "Early June peas. Packed by H. P. Cannon, at Bridgeville, Sussex Co., Delaware. Packed where grown."

The contents of this can were covered with a mucilaginous substance. Preservatives, copper, zinc, and lead were absent. There was a trace of dissolved tin.

No. 10902. Small peas, sifted. Markell Brothers, Baltimore. This sample was bought from C. I. Kellogg, 602 Ninth street, NW., and cost 15 cents. The label read: "Small peas, sifted; first quality. Packed at Baltimore, Md. Guaranteed finest quality. Markell Brothers."

The contents of the can were covered with a mucilaginous substance. No preservative was found. The peas were large and yellow. Copper was present, 15.1 mg per kilo being found. Zinc was absent, but tin and lead occurred, the former being in the larger amount.

No. 10903. Pois moyens. Brard & Cocary, Lorient. Bought of J. F. Page, 1210 F street NW., at a price of 20 cents. The label, beside the customary list of prizes and awards, read: "Pois moyens. Brard & Cocary. Maison à Lorient."

No preservative was found. The peas were large and green, and contained 17.1 mg of copper per kilo. There was no lead or zinc, but dissolved tin was present.

No. 10904. Petits pois extra fins. H. Lambert et Cie, Bordeaux. This sample was bought of J. F. Page, 1210 F street NW., at a price of 25 cents. It was labeled: "Petits pois extra fins, Henri Lambert et Cie, Bordeaux, France."

No preservative was found. The peas were small and green, and contained 65.8 mg of copper per kilo. There was no zinc, but tin and lead were present, the former in large proportions.

No. 10905. Napoleon brand marrowfat peas. F. H. Leggett & Co., New York. This sample was bought from J. F. Page, 1210 F street NW., and cost 13 cents. The label read: "Napoleon brand marrowfat peas. Francis H. Leggett & Co., West Broadway, Franklin and Varick streets, New York."

Salicylic acid was present in large quantity. The peas were large and yellow. Copper, lead and zinc were absent.

No. 10906. Choice green peas. Nail City Packing Co., Wheeling. This sample was bought from J. F. Page, 1210 F street NW., and cost 12¼ cents a can. The label was: "Choice green peas, Nail City Packing Co., Wheeling, W. Va."

No preservative could be detected. The peas had a slimy appearance, were large and somewhat soft. No copper was found, but zinc to the extent of 23.6 mg per kilo was present. There was a trace of lead, but no tin.

No. 10907. Petits pois extra fins. L. A. Price, Bordeaux. This sample was bought of G. E. Kennedy & Co., 1209 F street NW., and cost 40 cents a bottle. It was labeled: "Petits pois extra fins. Produits supérieure. L. A. Price, Bordeaux, France."

No preservative was found. The peas were small and green colored and had been coppered. The amount of the metal found was 28.3 mg per kilo. There was no zinc or lead.

No. 10980. Clymer early June peas. Mound City Preserving Co., St. Louis. This sample was bought from Schuster & Knox, Schuyler, Nebr., and cost 12½ cents. It was labeled "Clymer extra early June peas. Mound City Preserving Co., St. Louis, Mo."

The peas were large, yellow, soft, and of a slimy appearance. They contained no preservative which could be detected, but there was a little copper, 6.4 mg. No zinc was found, nor was lead. Tin to the amount of 23.9 mg per kilo was present.

No. 10981. White marrowfat peas. Curtice Bros. Co., Rochester, N. Y. This sample was bought of O. Nelson, Schuyler, Nebr., and cost 30 cents. The label read: "White marrowfat peas. Curtice Brothers Co., Rochester, N. Y., U. S. A. We guarantee the contents of this can to be of extra quality and packed at Rochester, Monroe Co., New York, U. S. A., Curtice Brothers Co., preservers. All goods under this label are of our own packing and are warranted to give entire satisfaction. Our fruits and vegetables are grown in this immediate vicinity especially for our wants."

No preservative was found. The peas were large and yellow, presented a slimy appearance and were somewhat soft. In spite of their color copper to the extent of 40 mg per kilo was present. There was no zinc or lead, but tin to the amount of 33.2 mg per kilo was detected.

No. 10983. White marrowfat peas. H. F. Hemingway & Co., Baltimore. This sample was bought from Towle & Morian, Schuyler, Nebr., and cost 15 cents. The label read: "White marrowfat peas. Packed by H. F. Hemingway & Co., Baltimore. First quality."

The peas were large and yellow, soft and slimy looking. No preservative or copper could be found. Lead and tin were present in about the proportions to form solder.

No. 10984. Morgan brand green peas. Cicero Canning Co., Chicago. This sample was bought from A. M. Parsons, Schuyler, Nebr., and cost 20 cents. It was labeled: "Green peas, the Morgan brand. Cicero Canning Co., Chicago, U. S. A."

Preservatives, copper, zinc, and lead were absent. Tin was present in solution. The peas were large, yellow, soft, and covered with a mucilaginous substance.

No. 10985. White marrowfat peas, Batavia Preserving Company, Batavia, N. Y. This sample was bought from A. M. Parsons, Schuyler, Nebr., and cost 20 cents. The label read: "White marrowfat peas. Batavia Preserving Co. Packed at Batavia, Genesee Co., N. Y. All goods bearing this trade mark are guaranteed to be of the finest quality. Packed by the Batavia Preserving Co., at Batavia, Genesee Co., N. Y. In packing the Batavia brand we employ the most experienced labor and have adopted the most modern machinery. The fruits are received daily from the growers, fresh from the field, consequently retain when packed all their natural flavor, as nothing but pure granulated sugar is used in the preserving process. The vegetables are packed with equal care, being brought daily from the field as rapidly as they have sufficiently matured thereby giving them freshness. After opening, empty contents of can into colander, straining off liquid before using."

The peas were large, yellow, hard, and apparently fresh. No preservatives were found. Copper, to the extent of 20.4 mg per kilo, was found.

Peas—Weights.

Serial No.	Price per package.	Weight of full package.	Weight of package.	Total contents.	Peas.	Weight of dry matter.	Dry matter.	Water.
	Cents.	Grams.	Grams.	Grams.	Grams.	Grams.	Per cent.	Per cent.
10625	20	707	122	585	....	84.2	14.40	85.60
10626	30	466	94	372	....	42.7	11.47	88.53
10627	20	716	135	581	....	85.7	14.75	85.25
10628	25	691	107	584	....	74.2	12.71	87.29
10629	30	492	98	394	....	47.5	12.05	87.95
10659	25	701	119	582	....	75.0	12.89	87.11
10661	30	493	87	406	....	43.1	10.62	89.38
10694	12½	716	134	588	490	117.8	20.03	79.97
10695	12	687	119	568	374	94.4	16.62	83.38
10696	12	711	121	590	385	102.8	17.43	82.57
10697	13	716	118	599	360	92.4	15.42	84.58
10698	14	690	119	571	401	90.8	17.48	82.53
10699	14	714	120	594	370	92.3	15.54	84.46
10700	11	698	117	581	363	89.4	15.38	84.62
10701	14	695	121	574	387	105.4	18.36	81.64
10702	10	704	124	580	401	74.2	12.80	87.20
10703	18	742	132	610	398	102.7	16.84	83.16
10704	10	690	130	560	540	125.5	22.41	77.59
10705	12	711	136	575	453	106.2	18.47	81.53
10706	15	685	119	566	385	94.7	16.74	83.26
10707	15	712	124	588	399	85.5	14.54	85.46
10708	13	702	130	572	353	89.5	15.64	84.36
10709	10	751	121	630	421	120.9	19.19	80.81
10710	15	705	128	577	375	97.1	16.49	83.51
10711	12	773	143	630	620	121.0	19.21	80.79
10712	10	723	124	599	360	96.9	16.17	83.83
10713	15	707	119	588	393	101.5	17.27	82.73
10714	18	686	117	569	351	62.2	10.94	89.06
10715	30	524	90	434	271	44.4	10.22	89.78
10716	25	530	93	437	283	57.8	13.23	86.77
10717	19	523	105	417	282	56.5	13.55	86.45
10718	20	496	100	396	268	57.3	14.46	85.54

WEIGHTS OF PEA SAMPLES.

*Peas—Weights—*Continued.

Serial No.	Price per package.	Weight of full package.	Weight of package.	Total contents.	Peas.	Weight of dry matter.	Dry matter.	Water.
	Cents.	Grams.	Grams.	Grams.	Grams.	Grams.	Per cent.	Per cent.
10719	15	531	114	418	272	49.1	11.74	88.26
10720	15	524	104	420	282	62.7	14.92	85.08
10721	30	524	93	431	293	42.7	9.90	90.10
10722	25	517	90	427	274	46.5	10.89	89.11
10723	25	490	85	405	309	48.8	12.04	87.96
10724	20	579	91	488	370	89.4	18.31	81.69
10870	15	515	107	408	280	61.1	14.97	85.03
10871	25	513	95	419	283	51.5	12.29	87.71
10872	18	511	105	406	266	49.3	12.15	87.95
10873	18	529	103	426	301	48.2	11.32	88.68
10874	15	527	89	438	289	67.2	15.34	84.66
10875	25	521	89	432	285	56.9	13.18	86.82
10876	12	510	94	416	283	67.2	16.16	83.84
10877	35	509	85	424	297	46.4	10.95	89.05
10878	20	523	86	437	285	77.8	17.80	82.20
10879	35	811	300	421	263	32.8	7.78	92.22
10880	35	517	99	418	313	41.7	9.98	90.02
10881	15	734	120	614	384	110.3	17.97	82.03
10882	12½	740	127	614	327	87.9	14.32	85.68
10883	18	731	126	605	390	100.7	16.65	83.35
10884	20	695	120	575	339	81.0	14.08	85.92
10885	40	800	352	448	303	44.3	9.89	90.11
10886	25	529	95	434	294	50.6	11.67	88.33
10887	25	522	99	423	308	63.5	15.02	84.98
10888	15	736	135	601	566	132.3	22.01	77.99
10889	16	514	85	429	290	69.8	16.27	83.73
10890	25	505	86	419	275	46.8	11.18	88.82
10891	30	509	100	409	294	45.6	11.15	88.85
10892	30	526	95	421	289	65.2	15.49	84.51
10893	16	520	99	421	301	65.6	15.59	84.41
10894	25	544	104	440	292	43.7	9.94	90.06
10895	20	550	104	446	299	42.9	9.62	90.38
10896	16	513	90	423	280	35.7	8.45	91.55
10897	35	754	325	429	260	36.3	8.46	91.54
10898	25	709	123	586	378	73.8	12.59	87.41
10899	30	710	116	594	393	65.8	11.07	88.93
10000	12½	706	130	576	352	41.8	7.26	92.74
10901	15	725	137	593	408	124.8	21.05	78.95
10002	15	707	125	582	364	87.9	15.10	84.90
10903	20	510	99	412	258	64.2	15.59	84.41
10904	25	510	85	413	284	40.0	9.69	90.31
10905	13	722	125	597	363	91.3	15.29	84.71
10906	12½	710	124	587	472	122.5	20.93	79.07
10907	40	781	350	431	283	44.5	10.33	89.67
10980	12½	715	119	596	373	109.1	18.31	81.69
10981	12½	722	131	641	474	124.3	19.39	80.61
10983	15	750	122	628	390	118.8	18.91	81.09
10984	20	746	147	590	553	135.0	22.54	77.46
10985	10	820	125	695	370	126.4	18.18	81.82

Peas.

Serial No.	Water.	Total dry matter.	Ether extract.	Crude fiber.	Ash.	Salt.	Corrected ash.	Nitrogen.	Albuminoids.	Digestible albuminoids.	Carbohydrates.
	Per ct.	Per ct.	Per ct.	Per ct.	Per ct.	Per ct.	Per ct.	Per ct.	Per ct.	Per ct.	Per ct.
10625	85.60	14.40	.442	1.228	.660		.241	.419	.559	3.489	3.288	8.581
10626	88.53	11.47	.195	.919	1.394		.644	.750	.380	2.375	1.857	6.587
10627	85.25	14.75	.215	1.081	1.127		.736	.391	.562	3.513	3.282	8.814
10628	87.29	12.71	.188	1.144	1.014		.636	.378	.534	3.340	3.012	7.024
10629	87.95	12.05	.219	1.202	.727		.310	.417	.512	3.200	3.011	6.703
10659	87.11	12.89	.205	1.177	.731		.322	.409	.579	3.619	3.256	7.158
10661	89.38	10.62	.137	.749	1.109		.768	.341	.461	2.881	2.757	5.744
10694	79.97	20.03	.208	1.126	.567		.024	.543	.795	4.969	4.312	13.160
10695	83.38	16.62	.203	1.171	1.218		.686	.532	.608	3.800	3.544	10.228
10696	82.57	17.43	.235	1.121	1.685	1.164	.521	.579	3.619	3.280	10.770
10697	84.58	15.42	.239	1.238	1.408		.948	.460	.577	3.606	3.306	8.929
10698	82.53	17.48	.267	1.381	1.332		.241	1.091	.604	4.338	3.964	10.162
10699	84.46	15.54	.205	1.265	1.455		.987	.468	.546	3.413	3.113	9.203
10700	84.62	15.38	.289	1.524	1.164		.675	.489	.618	3.803	3.505	8.540
10701	81.64	18.36	.301	1.324	1.439		.905	.534	.599	3.744	3.448	11.552
10702	87.20	12.80	.241	1.157	1.355		.937	.418	.490	3.063	2.781	6.981
10703	83.16	16.84	.253	1.465	1.206		.628	.578	.542	3.388	2.950	10.538
10704	77.59	22.41	.303	1.277	1.414		.840	.574	.892	5.575	5.040	13.841
10705	81.53	18.47	.281	1.094	.826		.323	.503	.700	4.375	3.971	11.894
10706	83.26	16.74	.291	1.250	.793		.288	.503	.658	4.113	3.611	10.203
10707	85.46	14.54	.237	1.176	.935		.438	.497	.571	3.569	3.117	8.623
10708	84.36	15.64	.280	1.334	.956		.491	.465	.624	3.900	3.544	9.170
10709	80.81	19.19	.359	1.341	1.128		.504	.564	.637	3.981	3.675	12.381
10710	83.51	16.49	.305	1.321	1.243		.947	.296	.510	3.188	2.749	10.433
10711	80.79	19.21	.254	1.385	.930		.354	.576	.672	4.200	3.861	12.441
10712	83.83	16.17	.285	1.015	1.505	1.119	.386	.534	3.338	3.066	10.027
10713	82.73	17.27	.347	1.285	1.154		.794	.360	.611	3.819	3.366	10.665
10714	89.06	10.94	.211	1.000	.926		.563	.363	.472	2.950	2.601	5.853
10715	89.78	10.22	.078	1.160	1.006		.724	.282	.380	2.375	2.183	5.601
10716	86.77	13.23	.069	1.283	1.058		.708	.350	.517	3.231	2.989	7.589
10717	86.45	13.55	.069	1.161	.865		.501	.364	.508	3.175	3.007	8.280
10718	85.54	14.46	.122	1.491	.596		.192	.404	.571	3.569	3.337	8.662
10719	88.26	11.74	.105	1.159	.984		.649	.335	.458	2.803	2.526	6.629
10720	85.08	14.92	.118	1.308	.947		.567	.380	.549	3.431	3.087	9.116
10721	90.10	9.90	.056	1.155	1.153		.866	.287	.396	2.475	2.210	5.002
10722	89.11	10.89	.142	1.158	1.044		.735	.309	.387	2.419	2.254	6.127
10723	87.96	12.04	.117	1.240	1.139		.701	.438	.543	3.394	3.085	6.150
10724	81.69	18.31	.203	1.434	1.011		.400	.551	.652	4.075	3.673	11.587
10870	85.03	14.97	.084	1.184	1.392		.494	.898	.501	3.094	3.367	8.612
10871	87.71	12.29	.232	1.124	1.080		.656	.424	.486	3.038	2.716	6.817
10872	87.95	12.15	.067	1.035	1.244		.874	.370	.459	2.860	2.477	6.935
10873	88.68	11.32	.207	.943	1.477	1.083	.394	.445	2.781	2.547	5.912
10874	84.66	15.34	.106	1.338	1.454	1.034	.420	.523	3.269	3.064	9.173
10875	86.82	13.18	.082	1.355	1.446	1.099	.347	.455	2.844	2.678	7.453
10876	83.84	16.16	.092	1.353	1.130		.640	.484	.561	3.506	3.255	10.070
10877	89.05	10.05	.204	1.043	1.246		.906	.340	.445	2.781	2.452	5.676
10878	82.20	17.80	.132	1.200	.990		.336	.654	.703	4.394	4.112	11.024
10879	92.22	7.78	.089	.556	.916		.603	.223	.293	1.831	1.584	4.388
10880	90.02	9.98	.202	.840	1.005		.640	.365	.447	2.794	2.650	5.139
10881	82.03	17.97	.428	1.236	1.438		.810	.619	.739	4.600	4.140	10.028
10882	85.68	14.32	.205	1.229	.752		.278	.474	.559	3.494	3.263	8.550

ANALYSES OF PEAS.

Peas—Continued.

Serial No.	Water.	Total dry matter.	Ether extract.	Crude fiber.	Ash.	Salt.	Corrected ash.	Nitrogen.	Albuminoids.	Digestible albuminoids.	Carbohydrates.
	Per ct.	Per ct.	Per ct.	Per ct.	Per ct.	Per ct.	Per ct.	Per ct.	Per ct.	Per ct.	Per ct.
10883	83.35	16.65	.313	1.380	1.129	.663	.466	.666	4.163	3.676	9.665
10884	85.92	14.08	.244	1.250	1.276	.794	.482	.591	3.694	3.475	7.610
10885	90.11	9.89	.175	.878	1.174	.920	.254	.381	2.381	2.122	5.282
10886	88.33	11.67	.099	1.083	1.458	1.127	.331	.434	2.713	2.383	6.317
10887	84.98	15.02	.125	1.514	1.084	.671	.413	.668	4.175	3.756	8.122
10888	77.99	22.01	.372	1.279	2.009	1.290	.710	.869	5.431	5.062	12.910
10889	83.73	16.27	.119	1.058	1.095	.514	.581	.675	4.219	3.871	9.779
10890	88.82	11.18	.142	1.195	.845	.493	.352	.468	2.925	2.637	6.063
10891	88.85	11.15	.226	1.130	.920	.564	.365	.498	3.113	2.801	5.752
10892	84.51	15.49	.136	1.293	.976	.514	.462	.626	3.913	3.572	9.173
10893	84.41	15.59	.273	1.381	.950	.539	.411	.658	4.113	3.748	8.875
10894	90.06	9.94	.212	.961	.519	.210	.309	.450	2.869	2.589	5.379
10895	90.38	9.62	.177	.888	.564	.250	.305	.447	2.794	2.555	5.197
10896	91.55	8.45	.154	.952	.341	.169	.172	.423	2.644	2.263	4.359
10897	91.54	8.46	.082	.737	.928	.638	.290	.361	2.256	1.822	4.457
10898	87.41	12.59	.218	1.190	.953	.533	.320	.551	3.444	3.112	6.785
10899	88.93	11.07	.183	1.003	.930	.496	.434	.547	3.419	3.085	5.475
10900	92.74	7.26	.106	.464	.604	.388	.216	.263	1.641	1.498	4.382
10901	78.95	21.05	.324	1.320	1.271	.562	.709	.808	5.050	4.564	13.075
10902	84.90	15.10	.221	1.147	1.643	1.128	.515	.631	3.944	3.520	8.145
10903	84.41	15.59	.158	1.249	1.433	.737	.696	.670	4.188	3.837	8.562
10904	90.31	9.69	.093	1.049	.860	.502	.305	.417	2.606	2.290	5.073
10905	84.71	15.29	.258	1.193	1.395	.914	.481	.616	3.850	3.410	8.594
10906	79.07	20.93	.316	1.421	1.482	.980	.502	.856	5.350	4.908	12.361
10907	89.67	10.33	.036	1.057	1.100	.853	.256	.407	2.544	2.369	5.584
10980	81.69	18.31	.353	1.285	1.298	.712	.586	.776	4.850	4.493	10.524
10981	80.61	19.30	.376	1.183	1.413	1.068	.345	.787	4.919	4.469	11.499
10983	81.09	18.91	.399	1.142	1.276	.834	.442	.749	4.681	4.272	11.412
10984	77.46	22.54	.401	1.267	1.413	.836	.577	.978	6.113	5.700	13.346
10985	81.82	18.18	.342	1.345	1.002	.649	.413	.809	5.038	4.601	10.393

FOODS AND FOOD ADULTERANTS.

Peas—Calculated to dry substance.

Serial No.	Ether extract.	Crude fiber.	Ash.	Salt.	Corrected ash.	Nitrogen.	Albuminoids.	Digestible albuminoids.	Per cent digestible.	Carbohydrates.
	Per ct.	Per ct.	Per ct.	Per ct.	Per ct.	Per ct.	Per ct.	Per ct.		Per ct.
10625	3.07	8.51	4.58	1.67	2.91	3.88	24.25	22.83	94.15	59.59
10626	1.70	8.01	12.18	5.61	6.57	3.31	20.69	16.19	78.25	57.42
10627	1.46	7.33	7.64	4.99	2.65	3.81	23.81	22.25	93.45	59.76
10628	1.48	9.00	7.98	5.00	2.98	4.20	26.25	23.70	90.29	55.29
10629	1.82	9.97	0.03	2.57	3.46	4.25	26.56	24.99	94.09	55.02
10659	1.59	9.13	5.67	2.50	3.17	4.49	28.06	25.26	90.02	55.55
10661	1.29	7.05	10.44	7.23	3.21	4.34	27.13	25.96	95.69	54.09
10694	1.04	5.62	2.83	.12	2.71	3.97	24.81	21.53	86.78	65.70
10695	1.22	7.05	7.33	4.13	3.20	3.66	22.88	21.33	93.22	61.52
10696	1.35	6.43	9.67	6.68	2.99	3.32	20.75	18.82	90.70	61.80
10697	1.55	8.03	9.13	6.15	2.98	3.74	23.38	21.44	91.70	57.91
10698	1.53	7.90	7.62	4.22	3.40	3.97	24.81	22.68	91.41	58.14
10699	1.32	8.14	9.36	6.35	3.01	3.51	21.94	20.03	91.29	59.24
10700	1.88	9.91	7.57	4.39	3.18	4.02	25.13	22.79	90.69	55.51
10701	1.64	7.21	7.84	4.93	2.91	3.26	20.38	18.78	92.15	62.93
10702	1.88	9.04	10.59	7.32	3.27	3.83	23.94	21.73	90.77	54.55
10703	1.50	8.68	7.16	3.73	3.43	3.22	20.13	17.52	87.03	62.53
10704	1.35	5.70	6.31	3.75	2.56	3.98	24.88	22.40	90.39	61.76
10705	1.52	5.92	4.47	1.75	2.72	3.79	23.69	21.50	90.76	64.40
10706	1.74	7.47	4.74	1.72	3.02	3.93	24.56	21.57	87.83	61.49
10707	1.63	8.09	6.43	3.01	3.42	3.93	24.56	21.44	87.30	59.29
10708	1.79	8.53	6.11	3.14	2.97	3.99	24.94	22.66	90.86	58.63
10709	1.87	6.99	5.88	2.94	2.94	3.32	20.75	19.15	92.29	64.51
10710	1.85	8.01	7.54	5.74	1.80	3.09	19.31	16.67	86.33	63.29
10711	1.32	7.21	4.84	1.84	3.00	3.50	21.88	20.10	91.86	64.75
10712	1.76	6.28	9.31	6.92	2.39	3.30	20.63	18.96	91.91	62.02
10713	2.01	7.44	6.68	4.60	2.08	3.54	22.13	19.49	88.07	61.74
10714	1.93	9.14	8.46	5.15	3.31	4.31	26.94	23.78	88.27	53.53
10715	.76	11.35	9.84	7.08	2.76	3.72	23.25	21.30	91.87	54.80
10716	.52	9.70	8.00	5.35	2.65	3.91	24.44	22.59	92.43	57.34
10717	.51	8.57	6.38	3.70	2.68	3.75	23.44	22.19	94.67	61.10
10718	.84	10.34	4.12	1.33	2.79	3.95	24.69	23.08	93.48	60.01
10719	.87	9.87	8.38	5.53	2.85	3.90	24.38	21.52	88.27	56.50
10720	.79	8.77	6.35	3.80	2.55	3.68	23.00	20.69	89.96	61.09
10721	.57	11.67	11.62	8.75	2.87	4.00	25.00	22.41	89.64	51.14
10722	1.30	10.63	9.59	6.75	2.84	3.55	22.19	20.70	93.28	56.29
10723	.97	10.30	9.46	5.82	3.64	4.51	28.19	25.62	90.88	51.08
10724	1.11	7.83	5.52	2.51	3.01	3.56	22.25	20.06	90.16	63.29
10870	.56	7.91	9.30	3.30	6.00	3.95	24.69	22.49	91.09	57.54
10871	1.89	9.15	8.79	5.34	3.45	3.95	24.69	22.10	89.51	55.48
10872	.55	8.52	10.24	7.19	3.05	3.78	23.63	20.39	86.29	57.06
10873	1.83	8.33	13.05	9.57	3.48	3.93	24.56	22.50	91.61	52.23
10874	.69	8.72	9.48	6.74	2.74	3.41	21.31	19.97	93.71	59.80
10875	.62	10.28	10.97	8.34	2.63	3.45	21.56	20.32	94.25	56.57
10876	.57	8.37	6.99	4.00	2.99	3.47	21.69	20.14	92.85	62.38
10877	1.86	9.53	11.38	8.27	3.11	4.06	25.38	22.39	88.22	51.85
10878	.74	7.08	5.56	1.89	3.67	3.95	24.69	23.10	93.56	61.93
10879	1.14	7.15	11.78	8.91	2.87	3.77	23.56	20.36	86.42	56.37
10880	2.02	8.42	10.07	0.41	3.60	4.48	28.00	26.44	94.43	51.49
10881	2.38	6.88	8.00	4.56	3.44	4.11	25.69	23.07	89.80	57.05
10882	2.00	8.58	5.25	1.94	3.31	3.90	24.38	22.70	93.48	59.73
10883	1.88	8.29	6.78	3.98	2.80	4.00	25.00	22.08	88.32	58.05
10884	1.73	8.88	9.06	5.64	3.42	4.20	26.25	24.68	94.02	54.08

LABELS OF HARICOT VERT SAMPLES.

Peas—Calculated to dry substance—Continued.

Serial No.	Ether extract.	Crude fiber.	Ash.	Salt.	Corrected ash.	Nitrogen.	Albuminoids.	Digestible albuminoids.	Per cent digestible.	Carbohydrates.
	Per ct.	Per ct.	Per ct.	Per ct.	Per ct.	Per ct.	Per ct.	Per ct.		Per ct.
10885	1.77	8.88	11.87	0.30	2.57	3.85	24.06	21.45	89.15	53.42
10886	.85	9.28	12.50	0.66	2.84	3.72	23.25	20.42	87.83	54.12
10887	.83	10.08	7.22	4.47	2.75	4.45	27.81	55.01	89.93	54.06
10888	1.09	5.81	9.13	5.90	3.23	3.95	24.09	23.00	93.16	58.68
10889	.73	6.50	6.73	3.16	3.57	4.15	25.94	23.79	91.71	60.10
10890	1.27	10.69	7.50	4.41	3.15	4.19	26.19	23.59	90.07	54.29
10891	2.03	10.13	8.33	5.06	3.27	4.47	27.94	25.12	89.91	51.57
10892	.88	8.35	6.30	2.32	2.98	4.04	25.25	23.06	91.33	59.22
10893	1.75	8.86	6.09	3.46	2.63	4.22	26.38	24.04	91.13	56.92
10894	2.13	9.07	5.22	2.11	3.11	4.62	28.88	26.05	90.20	54.10
10895	1.84	9.23	5.86	2.69	3.17	4.65	29.06	26.56	91.40	54.01
10896	1.82	11.26	4.03	.20	3.83	5.00	31.25	26.78	85.10	51.64
10897	.97	8.71	10.97	7.54	3.43	4.27	26.69	21.54	80.71	52.06
10898	1.73	9.45	7.57	4.23	3.34	4.38	27.38	24.72	90.29	53.87
10899	1.65	9.60	8.40	4.48	3.92	4.94	30.88	27.87	90.25	49.47
10900	2.25	6.40	8.32	5.34	2.98	3.62	22.63	20.64	91.21	60.40
10901	1.54	6.32	6.04	2.67	3.37	3.84	24.00	21.68	90.33	62.10
10902	1.46	7.00	10.88	7.47	3.41	4.18	26.13	23.31	89.21	53.93
10903	1.01	8.01	9.19	4.73	4.46	4.30	26.88	24.61	91.55	54.91
10904	.96	10.83	8.97	5.80	3.17	4.30	26.88	23.63	87.91	52.36
10905	1.69	7.80	9.12	5.98	3.14	4.03	25.19	22.30	88.53	56.20
10906	1.51	6.79	7.08	4.68	2.40	4.09	25.56	23.45	91.74	59.06
10907	.35	10.23	10.74	8.26	2.48	3.94	24.63	22.93	93.10	54.05
10980	1.93	7.02	7.09	3.89	3.20	4.24	26.50	24.54	92.60	57.46
10981	1.94	6.10	7.32	5.51	1.81	4.06	25.38	23.05	90.82	59.26
10983	2.11	6.04	6.75	4.41	2.34	3.96	24.75	22.59	91.27	60.35
10984	1.78	5.62	6.27	3.71	2.56	4.34	27.13	25.29	93.22	59.20
10985	1.88	7.40	5.84	3.57	2.27	4.45	27.81	25.31	91.01	57.07

"HARICOTS VERTS."

There were seven samples marked "haricots verts" examined, all being of French origin. All contained salicylic acid, and three contained copper, the quantity in two being over 18 mg per kilo.

DESCRIPTION OF SAMPLES.

No. 10738. Haricots verts extra, Dandicolle & Gaudin, Bordeaux. This sample was purchased from J. B. Bryan & Bro., 608 Pennsylvania avenue NW., and cost 40 cents. It was contained in a glass bottle with a lead top. The label read: "Haricots verts, extra. Dandicolle & Gaudin, 'limited,' Bordeaux, France."

The contents of the bottle were green and contained salicylic acid. No copper was found. Lead was not determined.

No. 10739. Haricots verts, extra fins, G. Talbot, Bordeaux. This sample was purchased from Frank Hume, 454 Pennsylvania avenue NW., and cost 30 cents. The label was: "Haricots verts, extra fins, Fabrique de conserves alimentaires, Bordeaux, an^{cne} Maison H. Bouillet & Talbot, G. Talbot, successeur."

1098   FOODS AND FOOD ADULTERANTS.

The can was corroded; the contents fresh and sweet and of a green color. Salicylic acid was present, but no copper, lead, or zinc.

No. 10936. Haricots verts, extra. E. Du Raix, Bordeaux. This sample was bought of John P. Love, 1534 Fourteenth street, and cost 35 cents. It was contained in a glass bottle with a lead top. The label was: "Haricots verts, extra. Eugène Du Raix, Bordeaux."

Salicylic acid was present, and also copper, but the quantity of this metal was small, being but 2.5 mg per kilo. This was probably accidentally present. Lead was not tested for.

No. 10938. Haricots verts, extra fins. A. Godillot, jne., Bordeaux. This sample was bought from J. H. Hungerford, 1334 Ninth street, and cost 35 cents. The label was: "Haricots verts, extra fins. Alexis Godillot, jne., Bordeaux."

The can was bright and clean. The beans were of a green color. Salicylic acid was present, but no copper.

No. 10939. Haricots verts. A. Godillot, jne., Bordeaux. This sample was bought from Elphonzo Youngs Co., 428 Ninth street NW., and cost 45 cents. It was in a glass bottle with a varnished tin top, the joint being made on a rubber ring. The label was: "Haricots verts. A. Godillot, jne., Bordeaux."

The beans were green in color. Salicylic acid had been used in preserving this sample, and also copper. The amount of the metal found was 21.2 mg per kilo, or 8.5 mg per bottle.

No. 10940. Haricots verts, surfins. Louit Frères & Co., Bordeaux. This sample was bought of Charles I. Kellogg, 602 Ninth street NW., and cost 10 cents. The label was: "Haricots verts, surfins. Louit Frères & Co., Bordeaux, 3 etablissements, St. Sernin, Turenne, Tivoli (au Bouscat)."

The can was bright and clean. The beans had a decided green color. Salicylic acid is used in this brand, but no copper.

No. 11214. Haricots verts, sur extra. Fontaine Frères, Paris. This sample was bought of G. G. Cornwell & Son, 1412 Pennsylvania avenue NW. Price 30 cents. Label: "Haricots verts, sur extra. Fontaine Frères, Paris; Société anonyme des champignonières de la Gironde."

Salicylic acid was found to be present. Copper was also found, and to the extent of 11.6 mg per kilo.

"Haricots verts"—Weights.

Serial No.	Price.	Weight of full package.	Weight of package.	Weight of beans.	Total contents.	Dry matter.	Dry matter.	Water.
	Cents.	Grams.	Grams.	Grams.	Grams.	Grams.	Per ct.	Per ct.
10738	40	820	426	214	394	17.6	4.46	95.54
10739	30	520	89	342	431	24.5	5.69	94.31
10936	35	774	381	349	393	19.7	5.02	94.98
10938	35	538	100	252	438	20.5	4.68	95.32
10939	45	735	334	216	401	15.5	3.87	96.13
10940	10	505	110	274	395	18.4	4.67	95.33
11214	30	528	51	308	477	25.1	5.27	94.73

LABELS OF STRING BEAN SAMPLES.

"Haricots verts."

Serial No.	Water.	Total dry matter.	Ether extract.	Crude fiber.	Ash.	Salt.	Corrected ash.	Nitrogen.	Albuminoids.	Digestible albuminoids.	Carbohydrates.
	Per ct.	Per ct.	Per ct.	Per ct.	Per ct.	Per ct.	Per ct.	Per ct.	Per ct.	Per ct.	Per ct.
10738	95.54	4.46	.033	.437	.058	.702	.296	.161	.006	.832	1.926
10739	94.31	5.69	.331	.529	.949	.509	.440	.217	1.356	.901	2.525
10936	94.98	5.02	.060	.514	1.180	.909	.271	.194	1.213	.967	2.053
10938	95.32	4.68	.042	.475	1.229	.895	.334	.165	1.031	.771	1.903
10939	96.13	3.87	.035	.435	.907	.697	.210	.147	.919	.657	1.574
10940	95.33	4.07	.087	.491	.896	.623	.273	.171	1.069	.765	2.127
11214	94.73	5.27	.016	.489	1.305	.979	.326	.185	1.156	.683	2.304

"Haricots verts"—Calculated to dry substance.

Serial No.	Ether extract.	Crude fiber.	Ash.	Salt.	Corrected ash.	Nitrogen.	Albuminoids.	Digestible albuminoids.	Per cent digestible.	Carbohydrates.
	Per ct.	Per ct.	Per ct.	Per ct.	Per ct.	Per ct.	Per ct.	Per ct.		Per ct.
10738	.74	9.80	23.71	17.09	6.62	3.62	22.63	18.65	82.41	43.12
10739	5.82	9.30	16.67	8.94	7.73	3.81	23.81	15.84	66.53	44.40
10936	1.20	10.24	23.51	18.12	5.39	3.86	24.13	19.27	79.86	40.92
10938	.90	10.14	26.26	19.13	7.13	3.52	22.00	16.47	74.86	40.70
10939	.91	11.24	23.44	18.00	5.44	3.81	23.81	16.98	71.31	40.60
10940	1.87	10.52	19.19	13.35	5.84	3.65	22.81	16.37	71.77	45.61
11214	.31	9.28	24.76	18.58	6.18	3.51	21.94	12.96	59.07	43.71

STRING BEANS.

Twenty samples of canned string beans were examined; fifteen were found to contain salicylic acid, and it was possibly present in two others. Three samples contained small amounts of copper and two contained a little zinc.

No. 10008. String beans. P. F. & D. E. Winebrenner, Baltimore, Md. This sample was bought from C. W. Proctor, corner of G and Thirteenth streets NW., and cost 20 cents. The label was: "String beans. Packed by P. F. & D. E. Winebrenner, at Baltimore, Md. First quality."

The can was clean and bright. A large amount of salicylic acid was found in the contents. A trace of lead was detected, but no copper or zinc.

No. 10660. Clipper brand string beans. Wm. Numsen & Sons, Baltimore. This sample was bought in Kissimmee, Fla., and cost 20 cents. The label was: "Clipper brand (established 1847) string beans. Packed by Wm. Numsen & Sons, at Baltimore, Md."

On opening the can there was a slight outflow of gas.

No. 10729. Maryland brand string beans T. J. Myer & Co., Baltimore. This sample was bought from J. B. Bryan & Bro., 608 Pennsylvania avenue, and cost 35 cents. It was labeled: "Maryland brand, string beans. First quality. Thos. J. Myer & Co., Baltimore, Md."

The can and contents were in good condition. Salicylic acid was present in large quantity, but copper and lead were absent. Zinc was found to the extent of 5.2 mg per kilo.

No. 10730. String beans. P. F. & D. E. Winebrenner, Baltimore. This sample was purchased from Frank Hume, 454 Pennsylvania avenue NW., and cost 10 cents. It was labeled. "String beans. Packed by P. F. & D. E. Winebrenner at Baltimore, Md."

The can and its contents were in good condition. A trace of salicylic acid was detected. Copper, lead and zinc were absent.

No. 10731. Choptank brand string beans. J. A. Wright & Bro., Choptank, Md. One can of this brand was bought from Beall & Baker, 486 Pennsylvania avenue NW., and cost 10 cents. Another can was procured from Frank Hume, 454 Pennsylvania avenue NW., at the same price. The label was: "Choptank brand string beans. Packed by J. A. Wright & Bro., Choptank, Md."

The can was not corroded. The contents were fresh and sweet. No preservative could be certainly identified, nor could copper be found.

No. 10732. Queen Anne brand string beans. J. F. Lowekamp, Jessups, Md. This sample was bought from Browning & Middleton, 610 Pennsylvania avenue NW., and cost 10 cents. It was labeled: "The 'Queen Anne' A. A. brand string beans. Best quality. Packed at Jessups, Anne Arundel Co., Md., by J. F. Lowekamp."

The can showed no corrosion. The contents were in good condition. Salicylic acid was present. There was no copper.

No. 10733. Standard string beans. Excelsior Canning Co., Maurertown, Va. This sample was bought from N. H. Shea, 632 Pennsylvania avenue, and cost 10 cents. The label was: "Standard string beans. Excelsior Canning Co., Maurertown, Va."

The can was clean and the contents fresh and sweet. Some salicylic acid was present, but copper could not be detected.

No. 10734. String beans. J. S. Farren, Baltimore. This sample was bought from Frank Hume, 454 Pennsylvania avenue NW., and cost 10 cents. It was labeled: "String beans. First quality. None genuine without the shield. Packed at Baltimore, Balto. Co., Md., by J. S. Farren & Co., Baltimore."

The can was clean and the contents were fresh and sweet. No salicylic acid or other antiseptic was found. Copper was also absent.

No. 10735. Blue Ridge string beans. B. F. Shriver & Co., Union Mills, Md. This sample was bought from A. A. Winfield, 215 Thirteen-and-a-half street SW., and cost 10 cents. The label was: "Blue Ridge brand string beans. Extra quality. B. F. Shriver & Co., Union Mills, Carroll Co., Md."

The can showed no corrosion. The contents were fresh and sweet. Some salicylic acid was present. No copper was found.

No. 10736. String beans. H. S. Lanfair & Co., Baltimore. This sample was bought from Estler Bros. & Co., 1301 C street SW., and cost 10 cents. It was labeled: "String beans. H. S. Lanfair & Co., Baltimore, Md. First quality. This can is the standard No. 2, adopted by the National Association of Canned Goods Makers. Packed at Baltimore, Md., by H. S. Lanfair."

The can and its contents were in good condition. No preservatives were found. A small amount of copper was found, 3.3 mg per kilo, or 1.9 mg per can. This was probably accidentally introduced.

LABELS OF STRING BEAN SAMPLES.   1101

No. 10737. Champion brand string beans. H. J. McGrath & Co., Baltimore. This sample was bought from S. S. Tucker, corner of C and Thirteenth streets SW., and cost 10 cents. The label was: "Champion brand string beans. First quality. Packed by H. J. McGrath & Co. at Baltimore, Md., U. S. A. Champion trade-mark established 1869."

The can and its contents were fresh, sweet, and clean. Salicylic acid was found, and also copper, the amount of the latter reaching 10.5 mg. per kilo (5.9 per can). This sample was a duplicate of No. 10927.

No. 10923. Choice string beans. E. B. Mallory & Co., Baltimore. Bought from W. T. Davis, 1467 P street NW., at a price of 15 cents. The label was: "Choice string beans, carefully packed at Baltimore, Md., by E. B. Mallory & Co."

The interior surface of the can showed no corrosion. A trace of salicylic acid was found. No copper was present.

No. 10924. String beans. Bamberger & Brewington, Baltimore. This sample was bought of M. F. Crown, 1532 Fourteenth street NW., and cost 15 cents. The full label was: "String beans. Packed by Bamberger & Brewington at Baltimore, Md."

The can was slightly corroded. Salicylic acid had been used to preserve this sample.

No. 10925. String beans. T. W. Bamberger & Co., Baltimore. This sample was bought from M. F. Crown, 1532 Fourteenth street NW., and cost 15 cents. The label was: "String beans. Packed by Thos. W. Bamberger & Co., Baltimore, Md." On the can was stamped: "Handmade. C. M. M. P. A."

The can was slightly corroded. Some salicylic acid had been used in this sample. No copper was present.

No. 10926. String beans. Githens & Rexsamer, Philadelphia. This sample came from John P. Love, 1534 Fourteenth street NW., and cost 20 cents. It was labeled: "Fresh string beans. These goods are of unsurpassed quality. Githens & Rexsamer, Philadelphia."

The inner surface of the can was found to be bright and clean. A small amount of salicylic acid was found in the contents, but no copper.

No. 10927. Champion brand string beans. H. J. McGrath & Co., Baltimore. This sample was bought from John W. Hardell, 1428 Ninth street NW., and cost 15 cents. The label was: "Champion brand string beans. First quality. Packed by H. J. McGrath & Co. at Baltimore, Md., U. S. A."

The inner surface of the can was bright and clean. A large amount of salicylic acid was present in the beans. No copper was present, but lead was found to the extent of 15.6 mg per kilo, possibly due, however, to comminuted solder.

No. 10928. October string beans. B. F. Shriver & Co., Union Mills, Md. This sample was bought from J. F. Page, 1210 F street NW., and cost 10 cents. The label was: "Blue Ridge October string beans, extra quality. B. F. Shriver & Co. Packed at Union Mills, Carroll Co., Md."

The interior surface of the can was bright and clean. Salicylic acid was found in the contents, but no copper could be detected.

No. 10929. Refugee string beans. Steele Brothers, New Britain, Conn. This sample was put up in a glass jar with a glass top, the joint being made on a rubber ring. It was bought from Geo. E. Kennedy & Co., 1209 F street NW., and cost 30 cents. The label was: "Refugee string beans. Put up by Steele Brothers, New Britain, Conn."

Salicylic acid was found in these beans in large quantity. This sample contained a little copper, the amount being 4.4 mg per kilo, or 3.3 mg per bottle. The quantity is hardly enough for coloring, and probably entered the beans accidentally. Lead was also present to the amount of 5.2 mg per kilo, or 3.9 mg per bottle.

No. 10930. Golden wax string beans. Steele Brothers, New Britain, Conn. This sample was bought from Geo. E. Kennedy & Co., 1209 F street NW., and cost 35 cents. The label was: "Golden wax string beans. Put up by Steele Brothers, New Britain, Conn."

The sample was contained in glass jars with glass tops, the joint being made on rubber bands. Salicylic acid was present in the contents. There was no copper present, but zinc was found to the extent of 3.2 mg per kilo, or 2.3 mg per bottle. Lead was likewise present, the amount being 34.4 mg per kilo or 24.8 mg per bottle. The sample used consisted of two bottles. The rubber ring from one was free from lead, though containing zinc, but the ring from the other contained 7.54 per cent of lead sulphate. It was probably from this source that the lead was absorbed by the sample.

No. 10990. Peerless string beans. C. H. Pearson Packing Co., Baltimore. This sample was bought from Towle & Morian, Schuyler, Nebr., and cost 15 cents. The label was: "Peerless brand string beans. Packed by the C. H. Pearson Packing Co., at Baltimore, Md."

The can was bright and clean. An enormous amount of salicylic acid was found in the contents, but there was no copper.

String beans—Weights.

Serial No.	Price.	Weight of full package.	Weight of package.	Weight of beans.	Total contents.	Dry matter.	Dry matter.	Water.
	Cents.	Grams.	Grams.	Grams.	Grams.	Grams.	Per cent.	Per cent.
10008	20	675	117	335	558	33.7	6.04	93.06
10729	35	3.375	347	1,785	3,028	222.0	7.33	92.67
10730	10	700	124	326	576	31.3	5.44	94.56
10731	10	665	119	321	546	28.5	5.22	94.78
10732	10	682	117	295	565	36.6	6.55	93.45
10733	10	633	117	336	516	47.7	9.25	90.75
10734	10	684	120	260	564	27.7	4.91	95.09
10735	10	700	120	293	580	29.4	5.07	94.93
10736	10	685	117	348	568	41.0	7.21	92.79
10737	10	677	115	335	562	36.8	6.55	93.45
10923	15	697	117	314	580	28.4	4.90	95.10
10924	15	687	127	247	560	27.1	4.87	95.13
10925	15	660	114	260	555	27.0	5.02	94.98
10926	20	666	123	243	543	22.7	4.17	95.83
10927	15	721	123	300	598	32.7	5.47	94.53
10928	30	701	128	254	573	31.1	5.43	94.57
10930	35	1.220	498	430	722	26.7	3.70	96.30
10990	15	699	120	300	560	27.7	4.04	95.06

ANALYSES OF STRING BEANS.

String beans.

Serial No.	Water.	Total dry matter.	Ether extract.	Crude fiber.	Ash.	Salt.	Corrected ash.	Nitrogen.	Albuminoids.	Digestible albuminoids.	Carbohydrates.
	Per ct.	Per ct.	Per ct.	Per ct.	Per ct.	Per ct.	Per ct.	Per ct.	Per ct.	Per ct.	Per ct.
10008	93.96	6.04	.066	.538	1.249	.909	.340	.148	.925	.657	3.262
10729	92.67	7.33	.095	.612	1.122	.394	.728	.182	1.138	.641	4.363
10730	94.56	5.44	.081	.513	1.132	.824	.308	.135	.844	.573	2.870
10731	94.78	5.22	.073	.636	.507	.204	.303	.158	.968	.698	3.016
10732	93.45	6.55	.082	.603	1.021	.655	.306	.144	.900	.540	3.944
10733	90.75	9.25	.130	.758	1.742	1.257	.485	.243	1.519	1.211	5.101
10734	95.09	4.91	.040	.393	1.335	.986	.349	.130	.813	.448	2.329
10735	94.93	5.07	.069	.471	1.391	.922	.469	.150	.994	.640	2.145
10736	92.79	7.21	.062	.632	.937	.567	.370	.177	1.106	.805	4.473
10737	93.45	6.55	.050	.562	.782	.411	.371	.171	1.069	.789	4.087
10923	95.10	4.90	.018	.439	2.258	1.988	.270	.102	.637	.332	1.548
10924	95.13	4.87	.044	.461	1.283	.999	.284	.124	.875	.471	2.307
10925	94.98	5.02	.046	.501	1.011	.689	.322	.145	.906	.622	2.556
10926	95.83	4.17	.039	.341	1.035	.746	.289	.107	.669	.361	2.086
10927	94.53	5.47	.054	.491	.946	.615	.331	.129	.806	.446	3.173
10928	94.57	5.43	.025	.471	1.360	.998	.362	.141	.881	.610	2.693
10930	96.30	3.70	.057	.439	.491	.216	.275	.120	.750	.560	1.969
10999	95.06	4.94	.037	.394	1.263	.960	.303	.115	.719	.484	2.527

String beans—Calculated to dry substance.

Serial No.	Ether extract.	Crude fiber.	Ash.	Salt.	Corrected ash.	Nitrogen.	Albuminoids.	Digestible albuminoids.	Per cent digestible.	Carbohydrates.
	Per ct.	Per ct.	Per ct.	Per ct.	Per ct.	Per ct.	Per ct.			Per ct.
10008	1.10	8.91	20.67	15.05	5.62	2.45	15.31	10.87	71.00	54.01
10660	.96	10.46	11.00	4.30	6.70	2.97	18.56	16.86	90.84	59.02
10729	1.30	8.35	15.31	5.38	9.93	2.48	15.50	8.74	56.39	59.54
10730	1.49	9.43	20.80	15.14	5.66	2.48	15.50	10.54	68.00	52.78
10731	1.40	12.18	9.71	3.90	5.81	3.02	18.88	13.38	70.87	57.83
10732	1.25	9.21	15.59	10.00	5.59	2.20	13.75	8.25	60.00	60.18
10733	1.41	8.19	18.83	13.59	5.24	2.63	16.44	13.09	79.62	55.13
10734	.81	8.01	27.19	20.08	7.11	2.65	16.56	9.13	55.13	47.43
10735	1.37	9.28	27.44	18.19	9.25	3.13	19.56	12.63	64.57	42.35
10736	.85	8.77	12.99	7.88	5.11	2.46	15.37	11.16	72.61	62.02
10737	.77	8.58	11.94	6.27	5.67	2.61	16.31	12.04	73.82	62.40
10923	.36	8.95	46.09	40.58	5.51	2.09	13.06	6.77	51.84	31.54
10924	.90	9.46	26.34	20.52	5.82	2.54	15.88	9.68	60.96	47.42
10925	.92	9.97	20.14	13.72	6.42	2.88	18.00	12.39	68.83	50.97
10926	.93	8.17	24.81	17.88	6.93	2.57	16.06	8.66	53.92	50.03
10927	.98	8.98	17.29	11.25	6.04	2.36	14.74	8.15	55.29	58.01
10928	.46	8.67	25.05	18.37	6.68	2.50	16.19	11.23	69.19	49.63
10930	1.53	11.86	13.26	5.83	7.43	3.23	20.19	15.14	75.00	53.16
10999	.75	7.98	25.56	19.44	6.12	2.33	14.56	9.79	67.24	51.15

STRINGLESS BEANS.

Six samples bearing this label were examined. All were of American origin. Five contained salicylic acid and it may have been present in the sixth, though it could not be identified with certainty. There was a trace of copper found in one sample.

DESCRIPTION OF SAMPLES.

No. 10740. Stringless beans. Curtice Brothers Co., Rochester, N. Y. This sample was bought from J. B. Bryan & Bro., 608 Pennsylvania avenue NW., and cost 30 cents. It was labeled: "Fine stringless beans, extra small. Curtice Brothers Co., Rochester, N. Y., U. S. A. All goods under this label are of our own packing and warranted to give entire satisfaction. We guarantee the contents of this can to be of extra quality and packed at Rochester, Monroe Co., New York, U. S. A. Our fruits and vegetables are grown in this immediate vicinity especially for our wants. Curtice Brothers Co., Preservers."

The can was not corroded and its contents were in good condition. Preservatives could not be detected with certainty. No copper was present.

No. 10931. Nonpareil stringless beans. W. L. Gardner, Jessups, Md. This sample was purchased of E. L. Yewell, 1141 Ninth street NW., and cost 10 cents. The label was: "Nonpareil stringless beans; extra quality. Packed by Wm. L. Gardner at Jessups, Howard County, Md."

The can was slightly corroded. The contents were well preserved, fresh, and sweet. A little salicylic acid was found to be present. A trace of copper (0.3 per kilo) was found, but it was undoubtedly present by accident.

No. 10932. Stringless beans. B. F. Shriver & Co., Union Mills, Md. This sample was bought from J. F. Russell, 730 Ninth street, and cost 17 cents. It was labeled: "Selected stringless beans, superlative quality. Packed by B. F. Shriver & Co. at Union Mills, Carroll County, Md."

The surface of the can was clean and bright, showing no corrosion. The contents were fresh and sweet. Salicylic acid had been used as a preservative. No copper or zinc was found, but lead to the extent of 28.4 per kilo was present. This lead possibly came from the solder.

No. 10933. Stringless beans. Thurber, Whyland & Co., New York. This sample was bought of Elphonzo Youngs Co., 428 Ninth street NW., and cost 18 cents. The label was: "Stringless beans, Thurber, Whyland & Co., New York. All goods bearing our name are guaranteed to be of superior quality, and dealers are authorized to refund the purchase price in any case where consumers have cause for dissatisfaction. It is therefore to the interest of both dealers and consumers to use Thurbers' brands."

The can was found to be clean and bright. The contents were sweet and fresh. Salicylic acid was present. No copper or zinc was found, but 18.0 mg of lead were found, due possibly to finely divided solder.

No. 10934. Mountain Rose stringless beans. Githens & Rexsamer, Philadelphia. This sample was bought from Elphonzo Youngs Co., 428 Ninth street, and cost 25 cents. The label was: "Mountain Rose stringless beans, Githens & Rexsamer, Philadelphia. These goods are of unsurpassed quality. Grown in a northern mountainous country, we have the result of delicacy, fine flavor, and extra tenderness. Careful heating only required to secure a delicious dish of beans."

The can was slightly corroded. The contents were fresh and sweet. Salicylic acid was present. No copper, lead, or zinc was present.

No. 10935. Stringless beans. Curtice Brothers Company, Rochester, N. Y. This sample was bought of Chas. I. Kellogg, 602 Ninth street NW., and cost 15 cents. It was labeled: "Stringless beans, Curtice Brothers Co.,

ANALYSES OF STRINGLESS BEANS.

•Rochester, N Y. All goods under this label are of our own packing and warranted to give entire satisfaction. We guarantee the contents of this can to be of extra quality and packed at Rochester, Monroe County, New York, U. S. A. Our fruits and vegetables are grown in this immediate vicinity especially for our wants. Curtice Brothers Co., Preservers."

The can was slightly corroded. The contents were in good condition. A little salicylic acid appeared to be present. No copper was found.

No. 11000. Stringless beans. *Mound City Preserving Company, St. Louis.* This sample was bought from O. Nelson, Schuyler, Nebr., and cost 15 cents. It was labeled: " Stringless beans. Mound City Preserving Co., St. Louis."

The can was bright and clean. The beans contained a small amount of salicylic acid and some sulphurous acid. No copper was present.

Stringless beans—Weights.

Serial No.	Price.	Weight of full package.	Weight of package.	Weight of beans.	Total contents.	Dry matter.	Dry matter.	Water.
	Cents.	Grams.	Grams.	Grams.	Grams.	Grams.	Per cent.	Per cent.
10740	30	686	111	356	575	34.2	5.95	94.05
10931	10	679	119	383	500	34.3	6.12	93.88
10932	17	701	121	329	580	33.0	5.69	94.31
10933	18	690	85	328	605	33.9	5.60	94.40
10934	25	695	123	....	572	32.3	5.65	94.35
10935	15	704	125	354	579	35.0	6.05	93.95
11000	15	707	114	416	593	45.1	7.60	92.40

Stringless beans.

Serial No.	Water.	Total dry matter.	Ether extract.	Crude fiber.	Ash.	Salt.	Corrected ash.	Nitrogen.	Albuminoids.	Digestible albuminoids.	Carbohydrates.
	Per ct.	Per ct.	Per ct.	Per ct.	Per ct.	Per ct.	Per ct.	Per ct.	Per ct.	Per ct.	Per ct.
10740	94.05	5.95	.096	.590	1.421	.935	.486	.203	1.269	.931	2.574
10931	93.88	6.12	.077	.624	1.199	.812	.387	.17\	1.069	.722	3.151
10932	94.31	5.69	.071	.446	1.450	1.085	.365	.156	.975	.664	2.748
10933	94.40	5.60	.066	.449	1.751	1.482	.269	.156	.975	.722	2.359
10934	94.35	5.65	.065	.561	.930	.482	.448	.184	1.150	.768	2.944
10935	93.95	6.05	.002	.577	1.677	1.192	.485	.188	1.175	.820	2.559
11000	92.40	7.60	.067	.776	1.395	.420	.975	.213	1.331	.684	4.031

Stringless beans—Calculated to dry substance.

Serial No.	Ether extract.	Crude fiber.	Ash.	Salt.	Corrected ash.	Nitrogen.	Albuminoids.	Digestible albuminoids.	Percent digestible.	Carbohydrates.
	Per ct.	Per ct.	Per ct.	Per ct.	Per ct.	Per ct.	Per ct.	Per ct.	Per ct.	Per ct.
10740	1.62	9.92	23.88	15.71	8.17	3.41	21.31	15.64	73.39	43.27
10931	1.26	10.20	19.59	13.26	6.33	2.79	17.44	11.79	67.71	51.51
10932	1.25	7.83	25.49	19.07	6.42	2.74	17.12	11.66	68.07	48.31
10933	1.18	8.02	31.26	26.47	4.79	2.78	17.38	12.90	74.22	42.16
10934	1.15	9.93	16.46	8.53	7.93	3.25	20.31	13.59	66.90	52.15
10935	1.03	9.53	27.72	19.71	8.01	3.10	19.38	13.56	69.97	42.34
11000	0.88	10.21	18.35	5.52	12.83	2.80	17.50	9.00	51.43	53.06

HARICOTS FLAGEOLETS.

Three samples of beans labeled "haricots flageolets" were examined. Two were packed in France and the third came from Great Britain. All were colored with copper. Salicylic acid was present in two. Lead was present in large quantity in two samples.

DESCRIPTION OF SAMPLES.

No. 10937. Haricots flageolets extra fins. E. Du Raix, Bordeaux. This sample was bought of John P. Love, 1534 Fourteenth street NW., and cost 35 cents. It was contained in a glass jar with a lead top. The label was: "Haricots flageolets, extra fins. Eugène Du Raix, Bordeaux."

The contents of the bottle were of a bright green color. A small amount of salicylic acid was detected. Copper was present to the extent of 25.2 mg per kilo or 10.7 mg per bottle. Lead was found to the enormous extent of 46.0 mg per kilo (19.6 mg per bottle). This was undoubtedly derived from the lead top.

No. 10941. Haricots flageolets extra fins. Amieux Frères, Paris. This sample was bought from Geo. E. Kennedy, 1209 F street NW., and cost 30 cents. The label was: "Haricots flageolets, fine fleur, extra fins. Amieux Frères, Paris, France. Se mefier des imitations. Notre devise est comme notre nom: Toujours à mieux."

The can was slightly corroded. The contents were of a deep green color and contained salicylic acid. Copper was present to the extent of 90.7 mg per kilo (39.8 mg per can). This is equivalent to 156.4 mg of copper sulphate in each can.

No. 10942. Flageolets extra fins. John Moir & Son, London. This sample was bought from A. O. Wright, 1632 Fourteenth street NW. Price, 50 cents. The label was: "Flageolets extra fins. Prepared by John Moir & Son, limited, London, Aberdeen, and Seville, purveyors to their royal highnesses the Prince of Wales, the Duke of Edinburgh, the Duke of Montpensier."

No preservatives were found. The beans, which were of a deep green color, were contained in a tall, slender bottle, closed at the top by a piece of varnished tinplate, the joint being made upon a disk of cork. Copper was present in this sample to the extent of 10.2 mg per kilo, or 5.2 mg per bottle. Lead was present in great quantity, there being 79.2 mg per kilo, or 40.7 mg per bottle. Its source is by no means clear. The disk of cork contained some lead, but this was more likely derived from the contents of the bottle than the reverse. The varnish on the tin top was lead-free. There was no zinc.

Luckily the price of this brand of goods is too high to permit its being in very general use.

LABEL OF HARICOT PANACHÉ SAMPLE.

Haricots flageolets—Weights.

Serial No.	Price.	Weight of full package.	Weight of package.	Solid contents.	Total contents.	Dry matter.	Dry matter.	Water.
	Cents.	Grams.	Grams.	Grams.	Grams.	Grams.	Per cent.	Per cent.
10937	35	797	372	283	425	83.3	19.60	80.40
10941	30	541	103	327	430	85.7	19.56	80.44
10942	50	1186	673	362	514	82.7	16.08	83.92

Haricots flageolets.

Serial No.	Water.	Total dry matter.	Ether extract.	Crude fiber.	Ash.	Salt.	Corrected ash.	Nitrogen.	Albuminoids.	Digestible albuminoids.	Carbohydrates.
	Per ct.	Per ct.	Per ct.	Per ct.	Per ct.	Per ct.	Per ct.	Per ct.	Per ct.	Per ct.	Per ct.
10937	80.40	19.60	.039	1.011	1.666	1.141	.525	.747	4.669	3.896	12.215
10941	80.44	19.56	.057	1.000	.882	.274	.608	.824	5.150	4.589	12.411
10942	83.92	16.08	.053	1.002	1.175	.727	.448	.645	4.031	3.253	9.819

Haricots flageolets—Calculated to dry substance.

Serial No.	Ether extract.	Crude fiber.	Ash.	Salt.	Corrected ash.	Nitrogen.	Albuminoids.	Digestible albuminoids.	Per cent digestible.	Carbohydrates.
	Per ct.	Per ct.	Per ct.	Per ct.	Per ct.	Per ct.	Per ct.	Per cent.		Per ct.
10937	.20	5.16	8.50	5.82	2.68	3.81	23.81	19.88	83.50	62.33
10941	.29	5.42	4.51	1.40	3.11	4.21	26.31	23.46	89.17	63.47
10942	.33	6.23	7.31	4.52	2.79	4.01	25.06	20.23	80.73	61.07

HARICOTS PANACHÉS.

But one brand of "haricots panachés" could be found on the Washington markets.

DESCRIPTION OF SAMPLE.

No. 10976. Haricots panachés. E. Du Raix, Bordeaux. This sample was bought from Elphonzo Youngs Co., 428 Ninth street NW., and cost 45 cents. It was labeled: "Haricots panachés, conserves extra. Eugène Du Raix, Bordeaux." The label was in red, white and blue, and represented a French flag.

This sample was put up in a glass bottle with a lead top, nothing intervening between this cover and the beans. There was a white coating of a lead salt on the inner side of the cover. The beans contained lead in solution, or at least in an oxidized state, to the amount of 15.6 mg per kilo, or 11.9 mg per bottle. Of course in this sample the occurrence of lead in the food can not be explained away as representing possible fragments of solder. Copper was also present to the amount of 30.4 mg per kilo. This is equivalent to 14.4 mg per bottle. Salicylic acid in an extremely large amount was also found. This combination of lead, copper and salicylic acid, all in one sample, must be trying to the stomach of the consumer.

The preserved foods of this packer and those of Dandicolle & Gaudin, also of Bordeaux, who use similar lead-topped bottles, should not be allowed on sale. It is unfortunate that they can not be excluded from the country.

Haricots panachés — Weights.

Serial No.	Price.	Weight of full package.	Weight of package.	Solid contents.	Total contents.	Dry matter.	Dry matter.	Water.
	Cents.	Grams.	Grams.	Grams.	Grams.	Grams.	Per cent.	Per cent.
10976	45	844	370	351	474	66.1	13.95	86.05

Haricots panachés.

Serial No.	Water.	Total dry matter.	Ether extract.	Crude fiber.	Ash.	Salt.	Corrected ash.	Nitrogen.	Albuminoids.	Digestible albuminoids.	Carbohydrates.
	Per ct.	Per ct.	Per ct.	Per ct.	Per ct.	Per ct.	Per ct.	Per ct.	Per ct.	Per ct.	Per ct.
10976	86.05	13.95	.028	1.014	1.007	.513	.494	.597	3.731	2.036	8.170

Haricots panachés—Calculated to dry substance.

Serial No.	Ether extract.	Crude fiber.	Ash.	Salt.	Corrected ash.	Nitrogen.	Albuminoids.	Digestible albuminoids.	Per cent digestible.	Carbohydrates.
	Per ct.	Per ct.	Per ct.	Per ct.	Per ct.	Per ct.	Per cent.	Per cent.		Per ct.
10976	.20	7.27	7.22	3.68	3.54	4.28	26.75	21.05	78.69	58.56

LITTLE GREEN BEANS.

One brand of goods bearing this label was found on the Washington markets.

DESCRIPTION OF SAMPLE.

No. 10944. Little green beans. Amherst Packing Co., North Amherst, Ohio. This sample was purchased of Chas. I. Kellogg, 602 Ninth street NW., and cost 10 cents. The label was: "The Amherst little green beans (stringless). The Amherst Packing Co. Packed at North Amherst, Lorain Co., O., U. S. A."

The can was slightly corroded. The contents were fresh and sweet. Salicylic acid was present. No copper was found, but there was a trace of lead, and 5.6 mg zinc per kilo or 3.3 mg per can.

Little green beans—Weights.

Serial No.	Price.	Weight of full package.	Weight of package.	Weight of solid contents.	Total contents.	Dry matter.	Dry matter.	Water.
	Cents.	Grams.	Grams.	Grams.	Grams.	Grams.	Per ct.	Per ct.
10944	10	713	131	360	582	36.0	6.18	93.82

ANALYSES OF WAX BEANS.

Little green beans.

Serial No.	Water.	Total dry matter.	Ether extract.	Crude fiber.	Ash.	Salt.	Corrected ash.	Nitrogen.	Albuminoids	Digestible albuminoids.	Carbohydrates.
	Per ct.	Per ct.	Per ct.	Per ct.	Per ct.	Per ct.	Per ct.	Per ct.	Per ct.	Per ct.	Per ct.
10944	93.82	6.18	.081	.048	1.502	1.070	.423	.184	1.150	.622	2.799

Little green beans—Calculated to dry substance.

Serial No.	Ether extract.	Crude fiber.	Ash.	Salt.	Corrected ash.	Nitrogen.	Albuminoids.	Digestible albuminoids.	Percent digestible.	Carbohydrates.
	Per ct.	Per ct.	Per ct.	Per ct.	Per ct.	Per ct.	Per cent.	Per cent.		Per ct.
10944	1.31	10.49	24.30	17.46	6.84	2.97	18.56	10.07	54.26	45.34

WAX BEANS.

One sample marked "wax beans" was examined.

DESCRIPTION OF SAMPLE.

No. 10943. Wax beans. Baker & Brown, Aberdeen, Md. This sample was bought from Charles I. Kellogg, 602 Ninth street NW., cost 15 cents, and was labeled: "Wax beans. Packed by Baker & Brown, Aberdeen, Harford Co., Md."

The can was clean and bright and the contents fresh. Salicylic acid was present and also zinc to the amount of 2.4 mg per can. Lead was also present. There was no copper.

Wax beans—Weights.

Serial No.	Price.	Weight of full package.	Weight of package.	Solid contents.	Total contents.	Dry matter.	Dry matter.	Water.
	Cents.	Grams.	Grams.	Grams.	Grams.	Grams.	Per cent.	Per cent.
10943	15	703	130	295	573	30.5	5.32	94.68

Wax beans.

Serial No.	Water.	Total dry matter.	Ether extracts.	Crude fiber.	Ash.	Salt.	Corrected ash.	Nitrogen.	Albuminoids.	Digestible albuminoids.	Carbohydrates.
	Per ct.	Per ct.	Per ct.	Per ct.	Per ct.	Per ct.	Per ct.	Per ct.	Per ct.	Per ct.	Per ct.
10943	94.68	5.32	.076	.578	1.184	.872	.312	.160	1.000	.761	2.482

Wax beans—Calculated to dry substance.

Serial No.	Ether extract.	Crude fiber.	Ash.	Salt.	Corrected ash.	Nitrogen.	Albuminoids.	Digestible albuminoids.	Per cent digestible.	Carbohydrates.
	Per ct.	Per ct.	Per ct.	Per ct.	Per ct.	Per ct.	Per cent.	Per cent.		Per ct.
10943	1.43	10.87	22.25	16.39	5.86	3.01	18.81	14.31	76.08	46.64

LIMA BEANS.

Fifteen samples of Lima beans were examined. Preservatives were found in ten cases. A little zinc was present in one sample and a little copper in another.

FOODS AND FOOD ADULTERANTS.

DESCRIPTION OF SAMPLES.

No. 10005. Lima beans. ———, *Baltimore.* This sample was bought from C. W. Proctor, corner of G and Thirteenth streets NW., and cost 10 cents. The label was partly illegible, and the packer's name could not be made out. The legible part read: "Lima beans; first quality. Packed at Baltimore."

The can was badly corroded. Salicylic acid was found in the contents. Copper to the extent of 5 mg per kilo was present, but no zinc. Lead to the extent of 10.5 mg per kilo was found.

No. 10741. Snowflake Lima beans. C. P. Mattocks, Portland, Me. One can of this brand was bought from J. J. Daly, 1367 C street SW., and another from N. H. Shea, 632 Pennsylvania avenue NW., each can costing 15 cents. The label was: "Snowflake Lima beans. Charles P. Mattocks, Portland, Me. Packed at Portland, Cumberland Co., Me. Extra quality. This can is packed from fresh beans. Every can guaranteed."

The can was much discolored and corroded. The contents were fresh and sweet. Sulphurous acid was found to be present. No copper, lead, or zinc could be detected.

No. 10742. Oval brand Lima beans. A. Booth Packing Company, Baltimore. This sample was bought from J. B. Bryan & Bro., 608 Pennsylvania avenue, and cost 10 cents. The label read: "Oval brand Lima beans. A. Booth Packing Co., Baltimore, Md., U. S. A. Packed at Baltimore, Baltimore Co., Md., U. S. A."

The can was much discolored and corroded. The contents were fresh and sweet. Sulphurous acid was present. There was no copper, lead, or zinc present.

No. 10743. Derby brand Lima beans. Wm. Numsen & Sons, Baltimore. This sample was purchased from J. B. Bryan & Bro., 608 Pennsylvania avenue, and cost 10 cents. It was labeled: "Derby brand Lima beans. Packed at Baltimore, Md., by Wm. Numsen & Sons."

The can was corroded and the contents slimy in appearance. Sulphurous acid and some salicylic acid were found. There was a trace of lead present, but no copper or zinc.

No. 10744. Lima beans. F. H. Leggett & Co., New York. This sample was bought from Jackson & Co., 626 Pennsylvania avenue NW., and cost 15 cents. The label read: "Lima beans. Francis H. Leggett & Co., West Broadway, Franklin and Varick streets, New York, N. Y. Packed at Baltimore, Md."

The can was corroded. The contents were fresh and sweet. No preservatives could be identified with certainty. There was a little zinc present (2 mg per kilo or 1.2 mg per can), but no lead or copper.

No. 10745. Equity brand Lima beans. Evans, Day & Co., Baltimore. This sample was bought from S. S. Tucker, corner of Thirteenth and C streets SW., and cost 13 cents. The label was: "Equity brand Lima beans. Packed by Evans, Day & Co., Baltimore City, Baltimore County, Md. First quality."

The can was much discolored and corroded. The contents were slimy. Sulphurous acid was present, but salicylic acid could not be identified with certainty. There was a trace of lead, but no copper or zinc.

No. 10746. Mountain Lima beans. Louis McMurray Packing Co., Frederick, Md. This sample was bought from Browning & Middleton, 610 Pennsylvania avenue NW., and cost 10 cents. The label was: "Mountain Lima beans; extra quality. Packed by the Louis McMurray Packing Co., successor to Louis McMurray, at Frederick City, Frederick County, Md. The seed for the Lima beans packed for this brand has been carefully selected. They are packed as soon as removed from the vines, thus insuring an excellent quality."

The can was much corroded and had become quite black. The contents appeared fresh and sweet. Sulphurous acid was present. There was no copper present, but there were found a trace of lead and 5.2 mg of zinc per kilo.

No. 10945. Onondaga cream Lima beans. Merrell & Soule, Syracuse. This sample was bought from E. E. Berry, stand 1, Riggs Market, and cost 15 cents. The label was: "Onondaga cream Lima beans. Merrell & Soule, Syracuse, N. Y. Indian brand, extra quality. Packed at Chittenango, Madison Co., N. Y."

The can was corroded. The contents were slimy in appearance. No antiseptic was found. No copper or zinc was found, but there was a trace of lead present.

No. 10946. Equity brand Lima beans. Evans, Day & Co., Baltimore. This sample was bought from F. L. Bubb, stand 64, Riggs Market, and cost 13 cents. The label was identical with that of 10745.

The can was corroded and the contents slimy. Both salicylic and sulphurous acids were present. No copper or zinc was found, but lead to the extent of 71.6 mg per kilo was found. This may have occurred as solder.

No. 10947. Preferred stock Lima beans. Maine State Packing Co., Portland, Me. This sample was bought from Birch & Co., 1414 Fourteenth street NW., and cost 18 cents. The label was: "Preferred stock Lima beans. Packed for the finest city trade by the Maine State Packing Co., Portland, Maine."

The can was badly corroded; the contents slimy. No preservative was found; no copper or zinc was present, but lead occurred to the extent of 56.0 mg per kilo. This may, of course, have existed in the state of solder.

No. 10948. California Lima beans. Los Angeles Packing Co. This sample was bought from John Keyworth, 318 Ninth street NW., and cost 15 cents. It was labeled: "California Lima beans. Los Angeles Packing Co."

The can was badly corroded. The beans were slimy and very dark. No preservative could be identified with certainty. No copper or zinc was present, but there was considerable lead (84.8 mg per kilo). This may, of course, have been present, at least partly, as solder.

No. 10949. Lima beans. Thurber, Whyland & Co., New York. This sample was bought from Elphonzo Youngs Co., 428 Ninth street NW., and cost 18 cents per can. The label was: "Lin . beans, Thurber, Whyland & Co.,

FOODS AND FOOD ADULTERANTS.

New York. Packed at Baltimore, Baltimore Co., Md. All goods bearing our name are guaranteed to be of superior quality and dealers are authorized to refund the purchase price in any case when consumers have cause for dissatisfaction. It is therefore to the interest of both dealers and consumers to use Thurbers' brands."

The can was slightly corroded. The contents were fresh and sweet. Some salicylic acid was present. No copper or zinc existed, but 33.2 mg per kilo of lead were found. Solder again may have been responsible.

No. 10995. Golden crown Lima beans. Aughinbaugh Canning Company, Baltimore. This sample was bought from Schuster & Knox, Schuyler, Nebr., and cost 13 cents. The label was: "Golden crown brand Lima beans. Packed by Aughinbaugh Canning Co., Baltimore, Md."

The can was slightly corroded. Sulphurous acid was present, but salicylic acid could not be certainly identified. No copper, lead or zinc was found.

No. 10996. Pride of California Lima beans. This sample was bought from Towle & Morian, Schuyler, Nebr., and cost 10 cents. The label as far as legible was: "Pride of California Lima beans. Natural flavor retained. Warranted first class."

The packer's name was torn off. The can was slightly corroded. A small amount of salicylic acid was found. No copper or zinc. Lead to the extent of 63.6 mg per kilo was found. It was probably due largely to finely divided solder.

No. 10997. Lima beans. Batavia Preserving Co., Batavia, N. Y. This sample was bought from A. M. Parsons, Schuyler, Nebr., and cost 20 cents. The label was: "Lima beans, packed by the Batavia Preserving Co., at Batavia, Genesee Co., N. Y., U. S. A. All goods bearing our trade mark are guaranteed to be of the finest quality."

The can was badly corroded. Salicylic acid was present There was no copper or zinc. Lead was found in enormous quantity, 97.6 mg per kilo. This was undoubtedly mostly due to solder.

Lima beans—Weights.

Serial No.	Price.	Weight of full can.	Weight of can.	Weight of beans.	Total contents.	Dry matter.	Dry matter.	Water.
	Cents.	Grams.	Grams.	Grams.	Grams.	Grams.	Per cent.	Per cent.
10005	10	719	132	429	587	98.3	16.74	83.26
10741	15	740	136	....	604	146.9	24.32	75.68
10742	10	748	115	359	623	102.0	16.12	83.88
10743	10	694	134	533	560	124.8	22.29	77.71
10744	15	702	128	352	574	93.8	16.34	83.66
10745	13	690	132	473	558	128.2	22.97	77.03
10746	10	688	115	381	573	112.8	19.60	80.39
10945	15	774	137	530	637	127.4	20.00	80.00
10946	12	692	131	400	561	118.7	21.16	78.84
10947	18	776	140	457	637	151.5	23.70	76.21
10948	15	789	148	577	641	151.3	23.61	76.39
10949	18	726	136	377	590	106.3	18.01	81.99
10995	13	722	135	335	587	96.7	16.47	83.53
10996	10	767	132	425	635	131.4	20.70	79.30
10997	20	737	139	....	598	135.4	22.65	77.35

LABELS OF BAKED BEAN SAMPLES.

Lima beans.

Serial No.	Water.	Total dry matter.	Ether extract.	Crude fiber.	Ash.	Salt.	Corrected ash.	Nitrogen.	Albuminoids.	Digestible albuminoids.	Carbohydrates.
	Per ct.	Per ct.	Per ct.	Per ct.	Per ct.	Per ct.	Per ct.	Per ct.	Per ct.	Per ct.	Per ct.
10005	83.26	16.74	.193	1.118	1.281	.586	.695	.628	3.925	3.107	10.223
10741	75.68	24.32	.598	1.394	1.248	.240	1.008	.732	4.575	3.998	16.505
10742	83.88	16.12	.287	.951	1.789	1.073	.716	.559	3.494	3.040	9.599
10743	77.71	22.29	.370	1.284	2.452	1.467	.985	.568	3.550	3.005	14.634
10744	83.66	16.34	.217	.904	1.569	.835	.734	.564	3.525	2.866	10.127
10745	77.03	22.97	.294	1.360	1.511	.508	1.003	.634	3.963	3.448	15.842
10746	80.39	19.61	.265	1.081	1.712	.816	.896	.616	3.850	3.257	12.702
10945	80.00	20.00	.278	1.022	1.592	.628	.964	.510	3.188	2.838	13.920
10946	78.84	21.16	.260	1.107	1.678	.703	.975	.536	3.350	3.037	14.765
10947	76.21	23.79	.307	1.311	2.550	1.694	.856	.766	4.788	4.368	14.744
10948	76.29	23.61	.210	1.263	1.768	.850	.918	.904	5.650	5.369	14.719
10949	81.99	18.01	.245	.994	1.162	.300	.862	.629	3.931	3.525	11.678
10995	83.53	16.47	.240	1.029	1.533	.805	.728	.507	3.169	2.699	10.499
10996	79.30	20.70	.232	1.151	1.430	.644	.786	.849	5.306	4.835	12.581
10997	77.35	22.65	.462	1.447	1.026	.387	.639	.591	3.694	3.219	16.021

Lima beans—Calculated to dry substance.

Serial No.	Ether extract.	Crude fiber.	Ash.	Salt.	Corrected ash.	Nitrogen.	Albuminoids.	Digestible albuminoids.	Percent digestible.	Carbohydrates.
	Per ct.	Per ct.	Per ct.	Per ct.	Per ct.	Per ct.	Per cent.	Per cent.		Per ct.
10005	1.15	6.08	7.05	3.50	4.15	3.75	23.44	18.56	79.18	61.08
10741	2.46	5.73	5.13	.97	4.16	3.01	18.81	16.44	87.40	67.87
10742	1.78	5.90	11.10	6.66	4.44	3.47	21.69	18.86	86.95	59.53
10743	1.66	5.76	11.00	6.58	4.42	2.55	15.94	13.48	84.56	65.64
10744	1.33	5.53	9.60	5.11	4.49	3.45	21.56	17.54	81.35	61.98
10745	1.28	5.92	6.43	2.21	4.22	2.76	17.25	15.01	87.01	69.12
10746	1.35	5.51	8.73	4.16	4.57	3.14	19.63	16.61	84.62	64.78
10945	1.39	5.11	7.96	3.14	4.82	2.55	15.94	14.19	89.02	69.60
10946	1.23	5.23	7.92	3.32	4.61	2.54	15.88	14.35	90.36	69.73
10947	1.07	5.51	10.72	7.12	3.60	3.22	20.13	18.36	91.21	61.97
10948	.89	5.35	7.49	3.60	3.89	3.83	23.94	22.74	94.99	62.33
10949	1.36	5.52	6.45	1.66	4.79	3.49	21.81	19.57	89.73	64.86
10995	1.46	6.25	9.31	4.89	4.42	3.08	19.25	16.39	85.14	63.73
10996	1.12	5.56	6.91	3.11	3.80	4.10	25.63	23.36	91.14	60.78
10997	2.04	6.39	4.53	1.71	2.82	2.61	16.31	14.21	87.12	70.73

BAKED BEANS.

Twelve samples of baked beans were examined, but several of these were duplicates of other samples. Salicylic acid certainly existed in 10 of these and probably in the remaining two. Sulphurous acid was found in three. Four samples contained copper and one also a trace of zinc.

DESCRIPTION OF SAMPLES.

No. 10006. Old South baked beans. Potter & Wrightington, Boston. This sample was bought from C. W. Proctor, corner of G and Thirteenth streets, NW., and cost 25 cents. The label read: "Old South brand, Boston baked beans. Packed and warranted by Potter & Wrightington, at Boston, Suffolk County, Mass., U. S. A. All goods bearing our signature are of first quality and are fully warranted in every respect."

Salicylic acid was present. No copper, lead, or zinc could be detected.

No. 10772. Picnic beans. Potter & Wrightington, Boston. This sample was bought from J. B. Bryan & Bro., 608 Pennsylvania avenue NW.; cost 10 cents, and was labeled: "Picnic beans. Old South brand Boston baked beans; first quality. Packed by Potter & Wrightington, Boston, Suffolk Co., Mass., U. S. A. All goods bearing our signature are of first quality, and are fully warranted in every respect. Potter & Wrightington. Put up expressly for use on the Pullman palace cars."

The can was corroded and colored black. The contents were fresh and sweet. Salicylic acid and sulphurous acid were both present. There was no copper or zinc, but lead was present.

No. 10772. Boston baked beans. C. Lewis & Co., Boston. This sample was bought from N. H. Shea, 632 Pennsylvania avenue NW., and cost 10 cents. It had evidently been in stock several years. The label was: "Boston baked beans; patented Aug., 1877. Put up and warranted by C. Lewis & Co., under the personal superintendence of W. K. Lewis, 426–428 Atlantic avenue, Boston." The rest of the label was illegible.

The can was corroded. The contents were fresh and sweet. A small amount of salicylic acid was detected. No copper or zinc, and but a trace of lead appeared to be present.

No. 10774. Boston baked beans. Grocers' Packing Co., Boston. This sample was bought from J. J. Daly, 1367 C street SW., and cost 15 cents. The label was: "Boston baked beans. Patented Aug. 7, 1877, W. K. Lewis, Boston. Packed and warranted by Grocer's Packing Co., Potter & Wrightington, prop's, Boston, Suffolk Co., Mass., U. S. A. Caution: None genuine without 'W. K. Lewis, Boston Baked Beans, patented Aug. 7, 1877,' stamped in the top of each can, with a label bearing the blue riband and seal of the United States. Under the patented process alone can the beans be warranted as perfect and to keep in any climate. Buyers are hereby cautioned, as all infringements are liable to seizure. Patent Office, United States of America."

The can bore all the insignia described. The contents were in good condition, but the can badly corroded. Salicylic acid and sulphurous acid were both found. This sample, like its duplicate, No. 10953, contained copper, the amount being 4 mg per kilo (4.4 per can). Lead was present, but zinc was absent.

No. 10775. Yankee baked beans. Curtice Brothers Co., Rochester, N. Y. This sample was bought from Jackson & Co., 626 Pennsylvania avenue NW., and cost 20 cents. The label was: "Yankee baked beans. Curtice Brothers Co., Rochester, N. Y., U. S. A. Extra quality. We guarantee the contents of this can to be of extra quality and packed at Rochester, Monroe Co., New York, U. S. A. Curtice Brothers Co., Preservers. All goods under this label are of our own packing and warranted to give entire satisfaction. Our fruits and vegetables are grown in this immediate vicinity, especially for our own wants. In the preparation of the Yankee baked beans we use only the best quality hand-picked pea beans and the choicest pig pork in proper proportions to secure that relishing and nutritious old-fashioned dish, so celebrated throughout the country."

The can was badly corroded. Salicylic acid was found in the beans and copper was also present. The quantity of the latter was large, being 88.9 mg per kilo. Lead was also present, but there was no zinc. There was no copper in a sample of the same brand (No. 11001) bought in Nebraska.

No. 10776. Old South Beans. Boston baked beans. Potter & Wrightington, Boston, Mass. One can of this sample was bought from Browning & Middleton, 610 Pennsylvania avenue, and one from Estler Bros. & Co., 1301 C street SW. Price of each, 15 cents. The label was: "Old South brand Boston baked beans. Packed and warranted by Potter & Wrightington, at Boston, Suffolk Co., Mass., U. S. A. All goods bearing our signature are first quality and are warranted in every respect. Potter & Wrightington."

"It is a thing all Yankees know
That beans with brown bread always go."

The can was badly corroded. No preservative could be identified with certainty. There was no copper in this sample, though it existed in other baked beans put up by what is apparently the same firm (Nos. 10953 and 10774). Neither was there copper, but lead to the extent of 186.0 mg per kilo was found. This was undoubtedly due to finely divided solder.

No. 10950. Boston baked beans. Amherst Packing Company, North Amherst, Ohio. This sample was bought from Hagan & Gilkison, stand 60, Riggs market, and cost 15 cents. The label was: "Boston baked beans, Amherst Packing Co., North Amherst, Ohio, U. S. A."

The can was slightly corroded. Salicylic acid was present. Copper was found to the extent of 6.6 mg per kilo, and zinc to the extent of 2.8.

No. 10951. Baked beans. Burnham & Morrill, Portland, Me. This sample was bought from C. F. Montgomery, 1506 Seventh street NW., and cost 15 cents. The label was: "Extra quality baked beans. Packed at Portland, Cumberland Co., Maine, U. S. A. Burnham & Morrill, Portland, Maine."

The can was slightly corroded. The beans contained no copper or zinc. Lead to the amount of 66.0 mg per kilo was found, possibly from finely divided solder. This sample showed evidence of containing a large amount of salicylic acid.

No. 10952. Baked beans. Wayne County Preserving Company, Newark, N. Y. This sample was bought from J. H. Hungerford, 1334 Ninth street NW., and cost 15 cents. The label was: "Extra quality baked beans, the Wayne County Preserving Co., Newark, Wayne County, N. Y. Established 1866. Packed at Newark, Wayne County, N. Y."

The can was badly corroded. Preservatives could not be certainly identified. There was no copper or zinc, but lead (69.2 mg per kilo) was present. This might have been in the form of solder, and very likely was, at least in part.

No. 10953. Boston baked beans. Grocers' Packing Company, Boston. This sample is a duplicate of No. 10774. It was bought from J. H. Hungerford, 1334 Ninth street NW., and cost 15 cents.

The can was badly corroded. Salicylic acid was present in the beans. There was some copper present, and in fairly large quantity, 7.5 mg per kilo (8 mg per can). It was also found in the duplicate sample.

Lead, to the extent of 72.0 mg per kilo, was also present, though possibly in the form of solder. There was no zinc.

No. 11001. Yankee baked beans. Curtice Brothers' Company, Rochester, N. Y. This sample is a duplicate of No. 10775. It was bought of O. Nelson, Schuyler, Nebr., and cost 25 cents.

Salicylic acid was found to be present. There was no copper or zinc, but a little lead (4.4 mg per kilo) was found.

No. 11002. The World's Fair baked beans. Royal Preserving Company, Chicago. This sample was bought in Schuyler, Nebr., of A. M. Parsons, and cost 20 cents. The label was: "1892. World's Fair baked beans. Westward the pride of Boston takes its way. Packed by the Royal Preserving Co., Chicago, U. S. A. Headquarters for pork, for beans, for cans, for boxes, for packing baked beans, and for the World's Fair."

The can was much discolored. The contents contained salicylic acid. No copper or zinc was found and only a trace of lead.

Baked beans—Weights.

Serial No.	Price.	Weight of full package.	Weight of package.	Weight of meat.	Total contents.	Dry matter.	Dry matter.	Water.
	Cents.	Grams.	Grams.	Grams.	Grams.	Grams.		Per cent.
10006	25	1,064	192	20	872	283.9	32.56	67.44
10772	10	390	91	....	299	213.1	35.66	64.34
10773	10	689	134	....	555	191.3	34.47	65.53
10774	15	1,308	188	35	1,120	363.1	32.49	67.51
10775	20	1,275	179	29	1,096	338.4	30.88	69.12
10776	15	1,269	188	48	1,081	160.9	33.43	66.57
10950	15	1,175	189	....	986	338.3	34.31	65.69
10951	15	1,141	183	36	958	293.2	30.61	69.39
10952	15	1,234	198	45	1,036	332.7	32.12	67.88
10953	15	1,267	197	30	1,070	358.7	33.52	66.48
11001	25	1,262	188	25	1,074	343.8	32.01	67.00
11002	20	1,215	172	50	1,043	329.9	31.63	68.37

Baked beans.

Serial No.	Water.	Total dry matter.	Ether extract.	Crude fiber.	Ash.	Salt.	Corrected ash.	Nitrogen.	Albuminoids.	Digestible albuminoids.	Carbohydrates.
	Per ct.	Per ct.	Per ct.	Per ct.	Per ct.	Per ct.	Per ct.	Per ct.	Per ct.	Per ct.	Per ct.
10006	67.44	32.56	1.273	1.332	2.400	1.322	1.087	1.224	7.650	6.404	19.896
10772	64.34	35.66	5.492	4.440	2.461	1.451	1.010	1.141	7.131	6.622	16.136
10773	65.53	34.47	4.884	1.375	1.723	.748	.975	1.182	7.338	7.056	19.150
10774	67.51	32.49	3.902	3.044	2.115	1.147	.968	1.069	6.681	6.251	16.743
10775	69.12	30.88	1.448	1.668	1.695	.840	.855	1.044	6.525	5.849	19.544
10776	66.57	33.43	2.410	2.180	2.150	.980	1.170	1.217	7.606	6.840	19.084
10950	65.69	34.31	4.134	3.246	2.625	1.071	1.554	1.136	7.100	6.210	17.205
10951	69.39	30.61	2.568	2.476	1.980	.928	1.052	1.022	6.388	5.528	17.198
10952	67.88	32.12	2.117	1.664	2.300	1.070	1.230	1.050	6.563	5.923	19.476
10953	66.48	33.52	4.596	3.748	2.360	1.227	1.133	1.200	7.500	6.553	15.316
11001	67.00	32.01	2.186	1.767	1.780	.707	.902	1.236	7.725	0.027	18.543
11002	68.37	31.63	3.119	2.550	2.141	.813	1.328	1.109	7.494	6.390	16.317

ANALYSES OF RED KIDNEY BEANS.

Baked beans—Calculated to dry substance.

Serial No.	Ether extract.	Crude fiber.	Ash.	Salt.	Corrected ash.	Nitrogen.	Albuminoids.	Digestible albuminoids.	Per ct. digestible.	Carbohydrates.
	Per ct.	Per ct.	Per ct.	Per ct.	Per ct.	Per ct.	Per ct.	Per ct.		Per ct.
10006	3.91	4.09	7.40	4.06	3.34	3.76	23.50	19.67	83.70	61.10
10772	15.40	12.45	6.90	4.07	2.83	3.20	20.00	18.57	92.85	45.25
10773	14.17	3.99	5.00	2.17	2.83	3.43	21.44	20.47	95.48	55.40
10774	12.01	9.37	6.51	3.53	2.98	3.29	20.56	19.24	93.58	51.55
10775	4.69	5.40	5.49	2.72	2.77	3.38	21.13	18.94	89.64	63.29
10776	7.21	6.52	6.43	2.93	3.50	3.64	22.75	20.46	89.94	57.08
10950	12.05	9.46	7.65	3.12	4.53	3.31	20.69	18.10	87.48	50.15
10951	8.39	8.09	6.47	3.03	3.44	3.34	20.88	18.06	86.49	56.17
10952	6.59	5.18	7.16	3.33	3.83	3.27	20.44	18.44	90.22	60.63
10953	13.74	11.18	7.04	3.66	3.38	3.58	22.38	19.55	87.36	45.66
11001	6.83	5.52	5.59	2.49	3.10	3.86	24.13	18.83	78.03	57.93
11002	9.86	8.09	6.77	2.57	4.20	3.79	23.69	20.23	85.39	51.59

RED KIDNEY BEANS.

But one sample of red kidney beans was examined. It was bought in Nebraska. No sample could be found in the Washington markets.

No. 10998. Red kidney beans. Batavia Preserving Company, Batavia, N. Y. This sample was bought from A. M. Parsons, Schuyler, Nebr., and cost 20 cents. The label read: "Red kidney beans. Batavia Preserving Company, Batavia, N. Y., U. S. A. All goods bearing this trade mark are guaranteed to be of the finest quality." On this label the words "extra stringless" have been crossed out and the words "red kidney" inserted with a hand stamp.

The can was badly corroded. Salicylic acid was found to be present in large quantities. There was no copper or zinc, but lead was present in large quantity, possibly, however, as solder.

Red kidney beans—Weight.

Serial No.	Price.	Weight of full package.	Weight of package.	Weight of solid contents.	Total contents.	Dry matter.	Dry matter.	Water.
	Cents.	Grams.	Grams.	Grams.	Grams.	Grams.	Per cent.	Per cent.
10988	20	744	147	573	597	163.2	27.33	72.67

Red kidney beans.

Serial No.	Water.	Total dry matter.	Ether extract.	Crude fiber.	Ash.	Salt.	Corrected ash.	Nitrogen.	Albuminoids.	Digestible albuminoids.	Carbohydrates.
	Per ct.	Per ct.	Per ct.	Per ct.	Per ct.	Per ct.	Per ct.	Per ct.	Per ct.	Per ct.	Per ct.
10998	72.67	27.33	.233	1.140	1.634	.700	.934	1.123	7.019	6.278	17.305

Red kidney beans—Calculated to dry substance.

Serial No.	Ether extract.	Crude fiber.	Ash.	Salt.	Corrected ash.	Nitrogen.	Albuminoids.	Digestible albuminoids.	Per cent digestible.	Carbohydrates.
	Per ct.	Per ct.	Per ct.	Per ct.	Per ct.	Per ct.	Per ct.	Per ct.		Per ct.
10098	.85	4.17	5.98	2.56	3.42	4.11	25.69	22.97	80.45	63.31

CORN.

Forty-one samples of canned corn were examined. Zinc occurred in 13; salicylic acid in 24. Copper was not found in any sample. Sulphurous acid occurred in many cases.

DESCRIPTION OF SAMPLES.

No. 10010. Susquehanna sugar corn. Mitchell Brothers, Aberdeen, Md. This sample was bought of C. W. Proctor, corner of Thirteenth and G streets NW., and cost 20 cents. The label read: "Susquehanna brand sugar corn. Packed by Mitchell Bros., at Aberdeen, Harford Co., Md."

The can was corroded. The corn was fresh and sweet. It contained 4.4 mg of zinc per kilo, but no copper. Lead (53.9 mg) was also found, but probably came from solder, at least in part.

No. 10749. Evergreen sugar corn. I. H. Houston, Vienna, Md. One can of this sample was bought from Jackson & Co., 626 Pennsylvania avenue NW.; one can from P. F. Bacon, 640 Pennsylvania avenue NW.; and another can from Frank Hume, 454 Pennsylvania avenue. Price of each can was 10 cents. The label read: "Evergreen sugar corn. Packed by I. H. Houston, at Vienna, Dorchester Co., Md. First quality, fine natural flavor."

The corn was white, soft and fresh. No preservative was detected. One can of this sample contained zinc to the amount of 73.2 mg per kilo; another, 42.6 mg. Lead in the first can was 23 mg per kilo, and in the other 42.2. Both may have been due to finely divided solder.

No. 10750. Nectarine sugar corn. B. F. Shriver & Co., Union Mills, Md. One can of this sample procured from N. H. Shea, 632 Pennsylvania avenue NW., cost 15 cents, and another, bought from Browning & Middleton, 610 Pennsylvania avenue, cost 12 cents. The label read: "Nectarine sugar corn, superlative. Packed by B. F. Shriver & Co., at Union Mills, Carroll County, Md."

The corn was white, fresh and soft. Salicylic acid was found in large amounts. Zinc was not present, but lead to the extent of 43.2 mg was found. This latter possibly was in the state of solder.

No. 10751. Pen-Mar brand sugar corn. John Root & Son, Mechanicstown, Md. One can of this sample was bought from Frank Hume, 454 Pennsylvania avenue NW., at a cost of 10 cents, and another was purchased from Beall & Baker, 486 Pennsylvania avenue NW., costing also 10 cents. The label read: "Pen-Mar brand sugar corn. First quality. Packed by John Root & Son, Mechanicstown, Frederick County, Md."

The corn was white, fresh and soft. No preservative could be detected. No zinc was found, but there was some lead (9.1 mg).

LABELS OF CORN SAMPLES. 1119

No. 10752. Premier corn. F. H. Leggett & Co. New York. This sample was bought from G. C. Burchard, 354 Pennsylvania avenue NW., and cost 15 cents. It was labeled: "Premier corn, Francis H. Leggett & Co., West Broadway, Franklin and Varick streets, New York."

The corn was white, fresh and soft. A large amount of salicylic acid was found. No zinc or lead was present.

No. 10753. Sugar corn. R. Williamson & Co., Baltimore, Md. One can of this brand was bought from S. S. Tucker, corner Thirteenth and C streets, SW., at a cost of 8 cents, and another from Frank Hume, 454 Pennsylvania avenue NW., for 10 cents. The label read: "Sugar corn, R. Williamson & Co's. brand. R. Williamson & Co., Baltimore, Md."

On a picture of an ear of corn appeared the words, "Soaked goods," in faint red letters. The corn was old, hard and yellow. No preservative was found. Zinc to a small amount (2.1 mg per kilo) was found, and there was also lead (34.8 mg). Possibly the lead was in the state of solder.

No. 10754. Egyptian sugar corn. C. A. McGaw, Perryman, Md. This sample was bought from Frank Hume, 454 Pennsylvania avenue NW., and cost 10 cents. It was labeled: "Egyptian sugar corn, first quality; choice packing. Packed where grown; extra quality. Packed by C. A. McGaw, at Perryman, Harford County, Md., expressly for the best family trade."

This corn was white, soft and sweet. It contained salicylic acid in large amount. There was a trace of lead, but no zinc.

No. 10755. Perfection sugar corn. J. C. Baker, Aberdeen, Md. This sample was bought from Frank Hume, 454 Pennsylvania avenue NW., and cost 10 cents a can. The label read: "J. C. Baker's perfection sugar corn; selected packing; first quality. Packed by J. C. Baker, at Aberdeen, Harford Co., Md."

The corn was white, soft and sweet. No preservative could be detected. There was no zinc present, but 66.8 mg lead were found. This may have been due partly to finely divided solder.

No. 10756. Friendship brand sugar corn. This sample was bought from Frank Hume, 454 Pennsylvania avenue, and cost 10 cents. The label read: "Friendship brand sugar corn, packed for family use; carefully selected, put up fresh, solid packed, and warranted." Packer's name illegible.

The corn was white, soft and sweet. In this sample salicylic acid was found. A large amount of zinc was also present, the quantity being 23.2 mg per kilo, or 13.9 mg per can. Lead was present in some quantity.

No. 10757. Snowflake sugar corn. C. P. Mattocks, Portland, Me. One can of this sample was bought from A. A. Winfield, 215 Thirteen-and-a-half street SW., and cost 15 cents, and another from Jackson & Co., 626 Pennsylvania avenue NW., at the same price. The can was labeled: "Snowflake sugar corn, Charles P. Mattocks, Portland, Me. This corn is packed in its own juices from selected ears. Every can guaranteed. Packed at Portland, Cumberland Co., Me. Extra quality."

The corn was white, soft and sweet. It contained salicylic acid. No zinc was found, but lead was present.

No. 10758. Blue Ridge brand sugar corn. B. F. Shriver & Co., Union Mills, Md. This sample was bought from J. J. Daly, 1367 C street SW., and cost 10 cents a can. The label read: "Blue Ridge brand, first quality sugar corn. Packed by B. F. Shriver & Co., at Union Mills, Carroll County, Md."

The corn was white, soft and sweet. Salicylic acid was found in large amount. Zinc was not detected, but lead was present.

No. 10759. Sugar corn. Nunley, Hynes & Co, Baltimore. This sample was bought from S. S. Tucker, corner of C and Thirteenth streets SW., and cost 8 cents. The label read: "Sugar corn. Nunley, Hynes & Co., Baltimore, Md."

The corn was hard, yellow and mature. This sample contained a large amount of salicylic acid. Zinc was present, 34.8 mg per kilo, or 19.8 per can, of the metal being found. There was also a trace of lead.

No. 10760. Sweet corn. Batavia Preserving Company, Batavia, N. Y. This sample was bought from Browning & Middleton, 610 Pennsylvania avenue NW., and cost 12 cents. The label read: "Sweet corn. Packed by the Batavia Preserving Co., at Batavia, Genesee Co., N. Y., U. S. A. All goods bearing this trade-mark are guaranteed to be of the finest quality."

The contents of the can were soft, mushy and white. A large amount of salicylic acid was found. No zinc was detected, but lead, possibly as solder, was present.

No. 10761. Swan Creek brand sugar corn. A. F. Brown, Havre de Grace. This sample was bought from A. A. Winfield, 215 Thirteen-and-a-half street SW., and cost 10 cents. It was labeled: "Swan Creek brand 1st quality sugar corn. Packed at Swan Creek, Harford Co., Md. Packed by A. F. Brown, Havre de Grace, Md."

The corn in this can was hard, fresh, sweet and white. No preservative was found, nor was zinc, but lead (9.2 mg per kilo) was present.

No. 10762. Gold leaf brand sugar corn. Western New York Preserving & Mfg. Co. Springville, N. Y. This sample was bought from A. A. Winfield, 215 Thirteen-and-a-half street SW. Price, 13 cents. The label read: "Gold leaf brand sugar corn. Packed at Springville, Erie Co., N. Y., by the Western New York Preserving & Mfg. Co."

The corn in this sample was soft, mushy and white. No preservative was found, nor was there zinc, but a trace of lead was detected.

No. 10763. Our choice corn. J. L. Barbour & Son, Washington. This sample was bought from S. S. Tucker, corner of C and Thirteenth streets SW., and cost 8 cents a can. The label read: "Our choice sweet corn. Packed for James L. Barbour & Son, Washington, D. C."

No preservatives were found in this sample. There was a small amount of zinc, 4.0 mg per kilo.

No. 10764. Sugar corn. C. G. Summers & Co., Baltimore. Bought from C. E. Nelson, corner of Seventh and I streets SE., at a price of 15 cents. Label: "Jas. K. Brown & Co. brand sugar corn. Soaked goods. Packed by Chas. G. Summers & Co., at Baltimore, Md."

The corn in this sample was hard, yellow and old. It was found to contain salicylic acid in large quantity. Zinc was also present to the extent of 2.8 mg per kilo. There was no lead.

LABELS OF CORN SAMPLES. 1121

No. 10765. Green corn. Baile & Stouffer, New Windsor, Md. This sample was bought from Frank Hume, 454 Pennsylvania avenue NW., and cost 10 cents. It was labeled: "Green corn. Packed by Baile & Stouffer, at New Windsor, Carroll Co., Md. All goods bearing this brand are guaranteed to be of first quality."

This corn was hard, white and fresh. It contained salicylic acid in large quantity, but no zinc. Lead was present.

No. 10766. Union Mills sugar corn. B. F. Shriver & Co., Union Mills, Md. This sample was bought from Frank Hume, 454 Pennsylvania avenue NW., and cost 10 cents. It was labeled: "Union Mills Packing Co.'s first quality sugar corn. Packed by B. F. Shriver & Co., at Union Mills, Carroll Co., Md."

The corn of this sample was soft, mushy and white. No preservative was found. Zinc and lead were also absent.

No. 10767. Boyle's Run brand sugar corn. G. H. Boyle, York Furnace, Pa. This sample was bought from H. I. Meader, 755 Eighth street SE., and cost 15 cents. The label read: "Boyle's Run brand sugar corn, first quality. Packed by Granville H. Boyle, York Furnace, York County, Pa."

The corn of this sample was soft, mushy and white. No preservative was found. Zinc to the amount of 6.4 mg per kilo and a trace of lead were present.

No. 10768. Western brand sugar corn. T. Clagett, Upper Marlboro, Md. This sample was bought from Frank Hume, 454 Pennsylvania avenue NW., and cost 10 cents. It was labeled: "Weston brand extra quality sugar corn. Packed by T. Clagett, near Upper Marlboro, Prince George Co., Md. This corn is grown and packed in Prince George Co., celebrated as the most fertile region of southern Maryland for the perfection of this vegetable."

No preservative was found in this brand. Zinc was present in small quantity (1.6 mg per kilo). There was no lead.

No. 10909. Rangeley sweet corn. A. H. Burnham, Waterford, Me. This sample was bought from E. E. Berry, stand 1, Riggs Market, at a cost of 15 cents. The label read: "Rangeley sweet corn. Packed by A. H. Burnham at Waterford, Oxford Co., Me. Selected and packed with especial care for finest quality Maine sugar corn."

The corn of this sample was soft, mushy and white. It contained both sulphurous and salicylic acids. Neither zinc nor lead was present.

No. 10910. Wakefield brand sugar corn. Smith, Yingling & Co., Westminster, Md. This sample was bought from J. R. Sherwood, stand 2, Riggs Market, and cost 10 cents. The label was: "Wakefield brand sugar corn; first quality. Packed where grown when perfectly fresh by Smith, Yingling & Co., at Westminster, Carroll Co., Md."

The can was corroded and the corn had a bad appearance. No preservative could be identified with certainty. Zinc was present. There was a trace of lead.

No. 10911. Scottish Chief sugar corn. Austin, Nichols & Co., New York. This sample was bought from F. E. Altemus, 1410 P street NW., and cost 12½ cents a can. The label bore a gaudy picture of a man dressed in full highland Scotch dress, and read: "Scottish Chief extra sugar corn.

"We guarantee all canned goods bearing our name to be of superior quality. Austin, Nichols & Co., Hudson, Jay, and Staples streets, New York. Packed at Taberg, Oneida Co., N. Y."

The corn was white. It contained both sulphurous and salicylic acids. There was no lead present, but zinc (3.2 mg per kilo), was found.

No. 10912. Preferred stock sugar corn. Maine State Packing Company, Portland, Me. This sample was bought from Birch & Co., 1414 Fourteenth street, and cost 18 cents. It was marked: "Preferred stock sugar corn; packed for finest city trade. Maine State Packing Co., Portland, Maine."

The corn was sweet and of a yellowish tint. Some salicylic acid was present, but neither lead nor zinc.

No. 10913. Monogram sugar corn. Githens & Rexsamer, Philadelphia. This sample was bought from M. F. Crown, 1532 Fourteenth street NW., and cost 15 cents. It was labeled: "Monogram brand crème de la crème sugar corn. Monogram sugar corn: The requisite, delicious, tender, sweet. Githens & Rexsamer, Philadelphia."

The inner surface of the can bore numerous black specks. The corn was white. Both salicylic and sulphurous acids were found. No zinc, and but a trace of lead, appeared to be present.

No. 10914. Creamlet sweet corn, Thurber, Whyland & Co., New York. This sample was bought from M. F. Crown, 1532 Fourteenth street NW., and cost 15 cents. The label read: "Creamlet sweet corn. Thurber, Whyland & Co., New York. All goods bearing our name are guaranteed to be of superior quality, and dealers are authorized to refund purchase price in any case where consumers have cause for dissatisfaction. It is therefore to the interest of both dealers and consumers to use Thurbers' brands. This corn is packed from the most tender and choice variety of green corn. It has already been thoroughly cooked, and only requires to be heated before serving on the table."

The can was corroded. The corn was fresh and sweet. Sulphurous acid was found. There was lead present (27.6 mg per kilo), but no zinc. The lead may have been due to solder.

No. 10915. Pen-Mar brand sugar corn. Root & Sons, Mechanicstown, Md. This sample was bought from John P. Love, 1534 Fourteenth street NW., and cost 10 cents. It is a duplicate of No. 10951. It was labeled: "Pen-Mar brand first quality sugar corn. Packed by John Root & Sons, Mechanicstown, Frederick Co., Md."

The can was corroded. The corn was fresh and sweet. Sulphurous acid was detected. Lead was present, but no zinc.

No. 10916. Honey-dew grated sugar corn. Erie Preserving Company, Buffalo. This sample was bought from John P. Love, 1534 Fourteenth street NW., and cost 25 cents. It was labeled: "Honey-dew grated sugar corn, Erie Preserving Co., Buffalo, Erie Co., N. Y., U. S. A. First quality. Packed at Brant, Erie Co., N. Y., U. S. A. Honey-dew brand, prepared solely for fancy trade with great care, only from the choicest and freshest vegetables obtainable. Each genuine can bears the signature of Erie Preserving Co., C. M. Fenton, Sec'y."

The can was corroded. The corn was fresh and sweet. Some salicylic acid was found. There was a trace of lead, but no zinc.

LABELS OF CORN SAMPLES.

No. 10917. Royal brand sugar corn. Northern Maine Packing Company, Dexter, Me. This sample was bought from John P. Love, 1534 Fourteenth street, and cost 15 cents. The label reads: "Royal brand, finest sugar corn; first quality. Packed at Dexter, Penobscot Co., Me., by the Northern Maine Packing Co."

The can was corroded. The corn was fresh and sweet. No preservative could be identified. Lead was present (55.2 mg per kilo) possibly as solder, but there was no zinc.

No. 10918. Kornlet. [Forestville Canning Co.] Forestville, N. Y. This sample was bought from A. O. Wright, 1632 Fourteenth street, and cost 25 cents a can. The label reads: "Kornlet; the choicest extract of green corn after the Forestville process. Fancy quality. Packed at Forestville, Chautauqua Co., N. Y., U. S. A."

The can was corroded. The corn was fresh and sweet. Sulphurous acid was present. No zinc was found, but there was a trace of lead.

No. 10919. Gaiety brand sugar corn. W. L. James, Hagerstown, Md. This sample was bought from H. Kengla, corner of Rhode Island avenue and Tenth street NW., and cost 7¼ cents a can. The label was: "Gaiety brand dry packed sugar corn. Grown in high latitudes from selected seed. Packed near Hagerstown, Washington Co., Md., by W. L. James."

The can was corroded. The corn was fresh and sweet. No preservative could be certainly identified. Zinc to the extent of 3.2 mg per kilo was present, and lead (28.8 mg) was also found. The latter, however, may have been present as solder.

No. 10920. Egyptian sugar corn. T. J. Myer & Co., Baltimore, Md. This sample was bought from J. H. Hungerford, 1334 Ninth street NW., at a cost of 10 cents. It was labeled: "Egyptian sugar corn; first quality. Packed by Thos. J. Myer & Co., at Baltimore, Md."

The can was corroded. The corn was fresh and sweet. Salicylic acid was detected. There was a trace of lead, but no zinc.

No. 10921. Bolling Brook sugar corn. C. K. Harrison, Upperville, Va. This sample was bought from J. H. Hungerford, 1334 Ninth street NW., and cost 10 cents. The label read: "Bolling Brook sugar corn. Chas. K. Harrison, Upperville, Va. Packed by the grower with much care, especially for family use."

The can was corroded. The corn was sweet and fresh. Both salicylic and sulphurous acids were found to be present, but neither lead nor zinc.

No. 10922. Honey-drop sugar corn. Davis, Baxter & Co., Portland, Me. This sample was bought from J. F. Russell, 730 Ninth street NW., and cost 15 cents. The label read: "Honey-drop sugar corn. Packed expressly for city trade by Davis, Baxter & Co., Portland, Me. New seed honey-drop. This corn is grown from seed selected and improved with the greatest care, and when it reaches the proper stage for picking it is cut from the cob and immediately sealed air-tight in its own milk. First quality. Packed at Winthrop, Kennebec County, Maine, U. S. A. See that label bears this signature: Davis, Baxter & Co."

The can was corroded. The corn was sweet and fresh. Both salicylic and sulphurous acids were detected. Lead (51.2 mg per kilo) was present, possibly in the shape of solder. No zinc was found.

No. 10987. Early sweet corn Curtice Brothers Co., Rochester, N. Y. This sample was bought from O. Nelson, Schuyler, Nebr., and cost 20 cents. The label read: "Early sweet corn, Curtice Brothers Co., Rochester, N. Y., U. S. A. All goods under this label are of our own packing and warranted to give entire satisfaction. We guarantee the contents of this can to be of extra quality and packed at Rochester, Monroe Co., New York, U. S. A. Curtice Brothers Co., preservers. Our fruits and vegetables are grown in this immediate vicinity, especially for our wants."

The can was slightly corroded. The presence of salicylic acid was detected. No lead or zinc was found.

No. 10989. Atlantic sugar corn. Atlantic Canning Company, Atlantic, Iowa. This sample was bought from O. Nelson, Schuyler, Nebr., and cost 15 cents. The label read: "Atlantic brand sugar corn. Packed by the Atlantic Canning Co., Atlantic, Iowa. All goods packed by us bearing our trademark are warranted to be first class."

The can was clean and bright. Salicylic acid was detected. No lead or zinc was found.

No. 10990. Standard sugar corn. Nebraska City Canning Company, Nebraska City, Nebr. This sample was bought from O. Nelson, Schuyler, Nebr., and cost 10 cents. It was labeled: "Standard sugar corn. Packed by the Nebraska City Canning Co., Nebraska City, Neb. Picked fresh from the farm. Fine quality. Natural flavor retained."

The can was corroded. The corn was fresh and sweet. Both salicylic and sulphurous acids were identified. No lead or zinc could be found.

No. 10991. Blair extra sugar corn. Blair Canning Company, Blair, Nebr. This sample was bought from Towle & Morian, Schuyler, Nebr., and cost 15 cents a can. The label was: "Blair extra sugar corn. Blair Canning Co., Blair, Nebraska."

The corn was fresh and sweet, and contained both sulphurous and salicylic acids. Zinc was also present in the large quantity of 20.0 mg per kilo or 12 mg per can. There was also a trace of lead.

No. 10993. White clover sugar corn. Bonney, Wheeler, Dingley & Co., Farmington, Me. This sample was bought from A. M. Parsons, Schuyler, Nebr., and cost 15 cents. The label read: "White clover sugar corn. This can is packed from selected corn. Every can guaranteed. Bonney, Wheeler, Dingley & Co., Farmington, Franklin Co., Maine."

The can was corroded, but the contents were remarkably clean and well preserved. Both salicylic and sulphurous acids were present. No lead or zinc was found.

No. 10994. Hawkeye brand sugar corn. Atlantic Canning Company, Atlantic, Iowa. This sample was bought from Bernard Mick, Schuyler, Nebr., and cost 15 cents. The label read: "Hawk-eye brand sugar corn. McWaid & Martin, Atlantic Canning Co., Atlantic, Iowa. All goods packed by us bearing our trade mark are warranted to be first class."

The can was corroded. The corn was mature and not inviting in appearance. Both salicylic and sulphurous acids were present. Lead was present to the extent of 20.8 mg per kilo, possibly as solder. Zinc was not found.

WEIGHTS OF CORN SAMPLES.

Corn—Weights.

Serial No.	Price.	Weight of full can.	Weight of can.	Total contents.	Dry matter.	Dry matter.	Water.
	Cents.	Grams.	Grams.	Grams.	Grams.	Per cent.	Per cent.
10010	20	704	118	586	129.7	22.13	77.87
10749	10	730	127	603	98.5	16.33	83.67
10750	12 and 15	714	128	586	123.6	21.09	78.91
10751	10	630	126	504	126.3	22.40	77.60
10752	15	756	136	620	167.5	27.01	72.99
10753	8 and 10	714	119	595	144.0	24.21	75.79
10754	10	738	136	602	144.7	24.04	75.96
10755	10	713	127	586	137.0	23.34	76.64
10756	10	731	130	601	158.7	26.41	73.59
10757	15	730	120	610	140.7	23.06	76.94
10758	10	744	122	622	147.0	23.63	76.37
10759	8	697	127	570	154.9	27.18	72.82
10760	12	745	127	618	173.7	28.10	71.90
10761	10	717	127	590	132.6	32.48	77.52
10762	13	760	127	633	143.6	22.69	77.31
10763	8	717	134	583	125.1	21.46	78.54
10764	15	700	115	585	133.1	22.76	77.24
10765	10	712	130	582	129.3	22.22	77.78
10766	10	767	134	633	186.4	29.39	70.61
10767	15	684	112	572	113.8	19.91	80.09
10768	10	762	127	635	133.6	21.04	78.96
10909	10	740	129	611	155.1	25.38	74.62
10910	10	714	126	588	139.0	23.64	76.36
10911	12½	712	125	587	141.5	24.11	75.89
10912	18	718	131	587	148.7	25.34	74.06
10913	15	747	145	602	122.3	20.32	79.68
10914	15	754	130	624	139.9	22.42	77.58
10915	10	689	127	562	140.3	24.96	75.04
10916	25	767	134	633	151.6	23.95	76.05
10917	15	805	144	661	209.4	31.68	68.32
10918	25	648	124	524	126.4	24.14	75.86
10919	7½	752	132	620	157.3	25.37	74.63
10920	10	734	124	620	164.5	26.53	73.47
10921	10	732	128	604	146.1	24.19	75.81
10922	15	756	140	616	180.4	29.28	70.72
10987	20	788	130	649	175.5	27.04	72.96
10989	15	740	120	620	146.6	23.64	76.36
10990	10	705	125	580	147.3	25.40	74.60
10991	15	730	119	611	153.8	25.18	74.82
10993	15	770	132	638	167.6	26.27	73.73
10994	15	752	125	627	160.0	25.52	74.48

1126   FOODS AND FOOD ADULTERANTS.

Corn.

Serial No.	Water.	Total dry matter.	Ether extracts.	Crude fiber.	Ash.	Salt.	Corrected ash.	Nitrogen.	Albuminoids.	Digestible albuminoids.	Carbohydrates.
	Per ct.	Per ct.	Per ct.	Per ct.	Per ct.	Per ct.	Per ct.	Per ct.	Per ct.	Per ct.	Per ct.
10010	77.87	22.13	1.328	.768	.724	.208	.516	.382	2.388	2.100	16.902
10749	83.67	16.33	.741	.532	.621	.186	.435	.315	1.969	1.737	12.467
10750	78.91	21.09	.787	.432	1.057	.529	.528	.388	2.425	1.997	16.389
10751	77.60	22.40	1.071	.616	1.210	.703	.507	.450	2.813	2.379	11.690
10752	72.99	27.01	1.396	.729	.675	.049	.626	.484	3.025	2.663	21.185
10753	75.79	24.21	1.566	.959	1.334	.922	.412	.446	2.788	2.559	17.563
10754	75.96	24.04	.959	.640	.664	.190	.474	.375	2.344	2.171	19.433
10755	76.64	23.34	1.284	.852	.649	.156	.493	.416	2.600	2.299	17.955
10756	73.59	26.41	1.144	.631	.903	.314	.589	.481	3.006	2.652	20.726
10757	76.94	23.06	1.303	.706	.625	.021	.604	.452	2.825	2.377	17.601
10758	76.37	23.63	1.063	.725	.763	.253	.510	.430	2.688	2.231	18.391
10759	72.82	27.18	1.894	1.242	1.487	.873	.614	.536	3.350	2.609	19.207
10760	71.90	28.10	1.281	.837	.767	.014	.753	.511	3.194	2.540	22.021
10761	77.52	22.48	1.209	.791	1.135	.688	.447	.362	2.263	1.967	17.082
10762	77.31	22.69	1.160	.860	.876	.290	.586	.495	3.094	2.634	16.691
10763	78.54	21.46	1.152	.818	1.116	.648	.468	.305	2.469	2.236	15.905
10764	77.24	22.76	1.347	.733	1.211	.772	.439	.453	2.831	2.333	16.638
10765	77.78	22.22	1.167	.756	1.255	.760	.495	.407	2.544	2.111	16.498
10766	70.61	29.39	1.837	1.149	1.393	.767	.626	.550	3.438	2.889	21.573
10767	80.09	19.91	.916	.657	.759	.283	.476	.380	2.375	1.939	15.203
10768	78.96	21.04	1.070	1.016	.747	.192	.555	.450	2.813	2.302	15.385
10909	74.62	25.38	1.393	.693	.680	.211	.469	.500	3.125	2.442	19.489
10910	70.67	29.33	1.560	.986	1.109	.443	.666	.499	3.129	2.795	22.540
10911	75.89	24.11	1.280	.709	.945	.328	.617	.523	3.270	2.879	17.906
10912	74.66	25.34	1.455	.798	1.021	.225	.796	.446	2.788	2.461	19.278
10913	79.68	20.32	1.329	.534	.679	.035	.644	.441	2.756	2.601	15.022
10914	77.58	22.42	1.134	.780	1.247	.531	.716	.442	2.763	2.551	16.496
10915	75.04	24.96	1.305	.921	1.233	.637	.596	.447	2.794	2.229	18.707
10916	76.05	23.95	.663	.412	.781	.259	.522	.359	2.244	1.614	19.850
10917	68.32	31.68	1.514	.710	.643	.029	.614	.596	3.725	2.091	25.088
10918	75.86	24.14	.850	.444	.857	.179	.678	.456	2.850	1.820	19.139
10919	74.63	25.37	1.195	.771	.751	.200	.551	.424	2.650	2.382	20.003
10920	73.47	26.53	1.555	1.157	1.398	.828	.570	.446	2.788	2.229	19.632
10921	75.81	24.19	1.239	.897	1.067	.567	.500	.409	2.556	1.888	18.431
10922	70.72	29.28	1.523	.861	.641	.056	.585	.515	3.219	2.375	23.036
10987	72.96	27.04	1.274	.838	.895	.287	.608	.484	3.025	2.636	21.008
10989	76.36	23.64	1.050	.934	.922	.291	.631	.456	2.850	2.428	17.884
10990	74.60	25.40	1.422	.922	1.344	.617	.727	.483	3.019	2.637	18.?
10991	74.82	25.18	1.246	.705	.886	.335	.551	.463	2.894	2.626	19.449
10992	71.87	28.13	1.516	1.150	.751	.118	.633	.591	3.694	3.328	21.010
10993	73.73	26.27	1.235	.828	.544	.034	.510	.431	2.694	2.469	20.969
10994	74.48	25.52	1.452	1.039	.689	.110	.579	.475	2.969	2.077	19.370

ANALYSES OF CORN.

Corn—Calculated to dry substance.

Serial No.	Ether extract.	Crude fiber.	Ash.	Salt.	Corrected ash.	Nitrogen.	Albuminoids.	Digestible albuminoids.	Per cent digestible.	Carbohydrates.
	Per ct.	Per ct.	Per ct.	Per ct.	Per ct.	Per ct.	Per ct.	Per ct.		Per ct.
10010	6.00	3.56	3.27	.94	2.33	1.73	10.81	9.49	87.70	76.36
10749	4.54	3.26	3.80	1.14	2.66	1.93	12.06	10.64	88.23	76.31
10750	3.73	2.05	5.01	2.51	2.50	1.84	11.50	9.47	82.35	77.71
10751	4.78	2.75	5.40	3.14	2.26	2.01	12.56	10.62	84.55	74.51
10752	5.17	2.70	2.50	.18	2.32	1.79	11.19	9.86	88.11	78.44
10753	6.47	3.96	5.51	3.81	1.70	1.84	11.50	10.57	91.91	72.56
10754	3.99	2.66	2.76	.79	1.97	1.56	9.75	9.03	92.62	80.84
10755	5.50	3.65	2.78	.67	2.11	1.78	11.13	9.85	88.50	76.94
10756	4.33	2.39	3.42	1.19	2.23	1.82	11.38	10.04	88.22	78.48
10757	5.05	3.06	2.71	.09	2.62	1.96	12.25	10.31	84.16	76.33
10758	4.50	3.07	3.23	1.07	2.16	1.82	11.38	9.44	82.95	77.82
10759	6.97	4.57	5.47	3.21	2.26	1.97	12.31	9.82	79.77	70.68
10760	4.56	2.98	2.73	.05	2.08	1.82	11.38	9.04	79.44	78.25
10761	5.38	3.52	5.05	3.06	1.99	1.61	16.09	8.75	86.07	75.90
10762	5.15	3.79	3.86	1.28	2.58	2.16	13.62	11.61	85.24	73.58
10763	5.37	3.81	5.20	3.02	2.18	1.84	11.50	10.42	90.61	74.12
10064	5.92	3.22	5.32	3.39	1.93	1.99	12.44	10.25	82.40	73.10
10765	5.25	3.40	5.65	3.42	2.23	1.83	11.44	9.50	83.04	74.26
10766	6.25	3.91	4.74	2.61	2.13	1.87	11.69	9.83	84.09	73.41
10767	4.60	3.30	3.81	1.42	2.39	1.91	11.94	9.74	81.57	76.35
10768	5.13	4.83	3.55	.91	2.64	2.14	13.38	10.94	81.76	72.11
10909	5.49	2.73	2.68	.83	1.85	1.97	12.31	9.02	78.15	76.79
10910	5.32	3.36	3.78	1.51	2.27	1.70	10.63	9.53	89.65	76.91
10911	5.31	2.94	3.92	1.36	2.56	2.16	13.50	11.94	88.44	74.33
10912	5.74	3.15	4.03	.85	3.18	1.76	11.00	9.71	88.27	76.06
10913	6.54	2.63	3.34	.17	3.17	2.17	13.56	12.80	94.40	73.93
10914	5.06	3.48	5.56	2.37	3.19	1.97	12.31	11.38	92.45	73.59
10915	5.23	3.09	4.94	2.55	2.39	1.79	11.19	8.93	74.85	74.95
10916	2.77	1.72	3.26	1.08	2.18	1.50	9.38	6.74	71.86	62.87
10917	4.78	2.24	2.03	.09	1.94	1.88	11.75	9.44	80.34	79.20
10918	3.52	1.84	3.55	.74	2.81	1.89	11.81	7.54	63.84	79.28
10919	4.71	3.04	2.96	.79	2.17	1.67	10.44	9.39	90.00	78.85
10920	5.86	4.36	5.27	3.12	2.15	1.68	10.50	8.40	80.00	74.01
10921	5.12	3.71	4.41	2.38	3.03	1.69	10.56	7.80	73.86	76.20
10922	5.20	2.94	2.19	.19	2.08	1.76	11.00	8.11	73.73	78.67
10987	4.71	3.10	3.31	1.06	2.25	1.79	11.19	9.75	87.13	77.69
10989	4.44	3.95	3.90	1.23	2.67	1.93	12.06	10.27	85.16	75.65
10990	5.60	3.63	5.29	2.43	2.86	1.90	11.88	10.38	87.37	73.60
10991	4.95	2.80	3.52	1.33	2.19	1.84	11.50	10.43	90.70	77.23
10992	5.39	4.09	2.67	.42	2.25	2.10	13.12	11.83	90.17	74.73
10993	4.70	3.15	2.07	.13	1.94	1.04	10.25	9.40	91.71	79.83
10994	5.69	4.07	2.70	.45	2.25	1.86	11.63	10.49	90.20	75.91

ARTICHOKE.

Three samples of artichokes were examined. Two were of French origin and one of American. Two samples contained copper and one salicylic acid. Zinc was not found.

DESCRIPTION OF SAMPLES.

No. 11215. Fonds d'artichauts. J. Nouvialle & Cie., Bordeaux. This sample was bought from G. G. Cornwell & Son, 1412 Pennsylvania avenue NW., and cost 75 cents. The label was: "Fonds d'artichauts. J. Nouvialle & Cie., Bordeaux, France."

No preservative was found in this brand, and there was no copper or zinc and but a trace of lead.

No. 11216. Fonds d'artichauts. Fontaine, Paris. This sample was bought from G. G. Cornwell & Son, 1412 Pennsylvania avenue NW., and cost 75 cents. The label read: "Fonds d'artichauts. Fontaine, Maison fondée in 1795. Usine à vapeur, 40 Boul'd National, Clichy, 'Seine.' R. du marche St. Honore, 14,16, et 18, Paris."

No preservative was found in this sample. Copper was found to the extent of 4.3 mg per kilo.

No. 11217. Green globe artichokes. G. W. Dunbar's Sons, New Orleans. This sample was likewise bought of G. G. Cornwell & Son, 1412 Pennsylvania avenue NW., and cost 45 cents. It was labeled: "Green globe artichokes. G. W. Dunbar's Sons, New Orleans. The heart of the artichaus is the most delicious and savory vegetable of the tropics. Canned au natural, they are sufficiently cooked to be eaten as they are, in salad or otherwise."

Salicylic acid was found in this artichoke sample. Copper to the amount of 6.6 mg per kilo was present. Lead (5.7 mg) was also found, but no zinc.

Artichokes—Weights.

Serial No.	Price.	Weight of full package.	Weight of package.	Solid contents.	Total contents.	Dry matter.	Dry matter.	Water.
	Cents.	Grams.	Grams.	Grams.	Grams.	Grams.	Per cent.	Per cent.
11215	75	492	99	248	393	38.5	9.79	90.21
11216	75	513	99	256	414	27.7	6.69	93.31
11217	45	569	113	226	456	28.0	6.15	93.85

Artichokes.

Serial No.	Water.	Total dry matter.	Ether extract.	Crude fiber.	Ash.	Salt.	Corrected ash.	Nitrogen.	Albuminoids.	Digestible albuminoids.	Carbohydrates.
	Per ct.	Per ct.	Per ct.	Per ct.	Per ct.	Per ct.	Per ct.	Per ct.	Per ct.	Per ct.	Per ct.
11215	90.21	9.79	.018	.604	2.229	1.795	.434	.123	.769	.528	6.170
11216	93.31	6.69	.009	.611	1.533	1.070	.463	.084	.525	.347	4.012
12217	93.85	6.15	.022	.534	1.400	.047	.453	.156	.075	.825	3.219

Artichokes—Calculated to dry substance.

Serial No.	Ether extract.	Crude fiber.	Ash.	Salt.	Corrected ash.	Nitrogen.	Albuminoids.	Digestible albuminoids.	Percent digestible.	Carbohydrates.
	Per ct.	Per ct.	Per ct.	Per ct.	Per ct.	Per ct.	Per cent.	Per cent.		Per ct.
11215	.18	0.17	22.77	18.33	4.44	1.26	7.88	5.39	68.40	69.00
11216	.13	0.13	22.02	15.99	6.93	1.26	7.88	5.19	65.86	59.94
11217	.36	8.08	22.76	15.40	7.36	2.53	15.81	13.42	84.88	52.30

SWEET POTATO.

But one brand of sweet potato was examined.

DESCRIPTION OF SAMPLE.

No. 11008. Baked sweet potato. Anderson Preserving Company, Camden, N. J. This sample was bought from A. M. Parsons, Schuyler, Nebr., and cost 20 cents. The label was: "New Jersey baked sweet potatoes. Packed by the Anderson Preserving Co., Camden Co., N. J."

The can was bright and clean. No preservative was found.

Sweet potatoes.

Serial No.	Price.	Weight of full package.	Weight of package.	Weight of total contents.	Dry matter.	Dry matter.	Water.
	Cents.	Grams.	Grams.	Grams.	Grams.	Per cent.	Per cent.
11008	20	1,165	187	978	309.4	31.64	68.36

Sweet potatoes.

Serial No.	Water.	Total dry matter.	Ether extract.	Crude fiber.	Ash.	Salt.	Corrected ash.	Nitrogen.	Albuminoids.	Digestible albuminoids.	Carbohydrates.
	Per ct.	Per ct.	Per ct.	Per ct.	Per ct.	Per ct.	Per ct.	Per ct.	Per ct.	Per ct.	Per ct.
11008	68.36	31.64	.256	.839	.842	.127	.715	.215	1.344	1.344	28.359

Sweet potatoes—Calculated to dry substance.

Serial No.	Ether extract.	Crude fiber.	Ash.	Salt.	Corrected ash.	Nitrogen.	Albuminoids.	Digestible albuminoids.	Per cent digestible.	Carbohydrates.
	Per ct.	Per ct.	Per ct.	Per ct.	Per ct.	Per ct.	Per ct.	Per ct.		Per ct.
11008	.81	2.65	2.66	.40	2.26	.68	4.25	4.25	100.00	89.63

OKRA.

Four samples of canned okra were purchased and examined. None of them contained zinc. Copper was present in one and salicylic acid in another.

DESCRIPTION OF SAMPLES.

No. 10769. Popular brand okra. Fait & Winebrenner, Baltimore. This sample was bought from Browning & Middleton, 610 Pennsylvania avenue NW., and cost 12 cents. The label was: "Popular brand okra. Fait & Winebrenner, Baltimore, Baltimore Co., Md."

The can was clean. No preservatives were found.

No. 10770. Leggett's dwarf okra. F. H. Leggett & Co., New York. This sample was bought from J. B. Bryan & Bro., 608 Pennsylvania avenue NW., and cost 15 cents. The label was: "Leggett's dwarf okra. Francis H. Leggett & Co., New York City. Packed at Riverside, Burlington County, New Jersey."

1130   FOODS AND FOOD ADULTERANTS.

The can was clean and bright. Salicylic acid was present.

No. 10972. Fresh okra. Githens, Rexsamer & Co., Philadelphia. This sample was bought from Elphonzo Youngs Co., 428 Ninth street NW., and cost 18 cents. The label was: "Fresh okra. Githens, Rexsamer & Co., Philadelphia. These goods are of unsurpassed quality."

The can was clean and bright. Preservatives were not found. Copper was found to the amount of 3.2 mg per kilo or 2.8 mg per can. This small amount may have been accidentally introduced. No zinc was present. Lead (18.8 mg per kilo) was also found, but may have been in the shape of solder.

No. 10973. Fresh dwarf okra, G. W. Dunbar's Sons, New Orleans. This sample was bought from Elphonzo Youngs Co., 428 Ninth street NW., and cost 18 cents. The label was: "Fresh dwarf okra. G. W. Dunbar's Sons, New Orleans, packers of semitropical products. All our goods are warranted first-class, full weight, and will keep in any climate."

The can was clean and bright. Preservatives were not found. No copper or zinc was obtained, but lead was present to the extent of 17.6 mg per kilo, possibly as solder.

Okra—Weights.

Serial No.	Price.	Weight of full can.	Weight of can.	Solid contents.	Total contents.	Dry matter.	Dry matter.	Water.
	Cents.	Grams.	Grams.	Grams.	Grams.	Grams.	Per cent.	Per cent.
10769	12	667	119	231	548	28.1	5.12	94.88
10770	15	680	125	445	555	32.9	5.92	94.08
10972	18	1,060	187	345	873	52.2	5.98	94.02
10973	18	1,089	178	893	911	50.7	5.57	94.43

Okra.

Serial No.	Water.	Total dry matter.	Ether ex. tract.	Crude fiber.	Ash.	Salt.	Corrected ash.	Nitrogen.	Albuminoids.	Digestible albuminoids.	Carbohydrates.
	Per ct.	Per ct.	Per ct.	Per ct.	Per ct.	Per ct.	Per ct.	Per ct.	Per ct.	Per ct.	Per ct.
10769	94.88	5.12	.242	1.373	.341	.040	.301	.107	.669	.360	2.495
10770	94.08	5.92	.038	.459	1.421	.972	.449	.144	.900	.346	3.102
10972	94.02	5.98	.048	.444	1.460	1.007	.453	.117	.731	.377	3.297
10973	94.43	5.57	.046	.352	1.701	1.300	.401	.086	.538	.168	2.933

Okra—Calculated to dry substance.

Serial No.	Ether ex. tract.	Crude fiber.	Ash.	Salts.	Corrected ash.	Nitrogen.	Albuminoids.	Digestible albuminoids.	Per cent digestible.	Carbohydrates.
	Per ct.	Per ct.	Per ct.	Per ct.	Per ct.	Per ct.	Per ct.	Per ct.		Per ct.
10769	4.73	26.81	6.66	.78	5.88	2.08	13.00	7.03	54.08	48.80
10770	.64	7.75	24.00	16.42	7.58	2.43	15.10	5.84	38.44	52.42
10972	.81	7.43	24.42	16.84	7.58	1.95	12.19	6.31	51.76	55.10
10973	.82	6.31	30.53	23.34	7.19	1.54	9.63	3.02	31.36	52.71

BRUSSELS SPROUTS.

But one brand of Brussels sprouts could be found on the Washington market. It was contained in a lead-topped bottle.

DESCRIPTION OF SAMPLE.

No. 10979. Choux de Bruxelles. Dandicolle & Gaudin, Bordeaux. This sample, which was contained in a glass bottle with a lead top, was bought from Elphonzo Youngs Co., 428 Ninth street NW., and cost 45 cents. It was labeled: "Choux de Bruxelles. Dandicolle & Gaudin, Bordeaux."

Salicylic acid was present. Copper was present to the enormous extent of 63.7 mg per kilo or 26.9 mg per bottle. Lead was not contained in this sample, but this must have been due to good luck, since there was nothing whatever in the way of an interposition between the lead of the cover and the food. Probably the sample kept right side up all the way during its long trip from Bordeaux to Washington. Not all these lead-topped samples met with the same good fortune, as may be seen by referring to No. 10937.

Brussels sprouts—Weights.

Serial No.	Price.	Weight of full package.	Weight of package.	Solid contents.	Total contents.	Dry matter.	Dry matter.	Water.
	Cents.	Grams.	Grams.	Grams.	Grams.	Grams.	Per ct.	Per ct.
10979	45	773	350	254	423	26.4	6.25	93.75

Brussels sprouts.

Serial No.	Water.	Total dry matter.	Ether extract.	Crude fiber.	Ash.	Salt.	Corrected ash.	Nitrogen.	Albuminoids.	Digestible albuminoids.	Carbohydrates.
	Per ct.	Per ct.	Per ct.	Per ct.	Per ct.	Per ct.	Per ct.	Per ct.	Per ct.	Per ct.	Per ct.
10979	93.75	6.25	.071	.502	1.273	.929	.344	.238	1.488	1.164	2.856

Brussels sprouts—Calculated to dry substance.

Serial No.	Ether extract.	Crude fiber.	Ash.	Salt.	Corrected ash.	Nitrogen.	Albuminoids.	Digestible albuminoids.	Percent digestible.	Carbohydrates.
	Per ct.	Per ct.	Per ct.	Per ct.	Per ct.	Per ct.	Per cent.	Per cent.		Per ct.
10979	1.14	8.99	20.37	14.87	5.50	3.80	23.75	18.63	78.44	45.75

TOMATOES.

Ten samples of tomatoes were examined and salicylic acid found in seven.

DESCRIPTION OF SAMPLES.

No. 10002. Cedar tomatoes. W. L. Stevens, Cedarville, N. J. This sample was bought from C. W. Proctor, corner of G and Thirteenth streets, NW., and cost 25 cents. The label read: "Cedar brand tomatoes; first quality. Packed by W. L. Stevens, Cedarville, Cumberland Co., New Jersey."

The can was stained a dark color on the interior. Salicylic acid was found in its contents.

No. 11003. Glenwood tomatoes. Glenwood Canning Company, Glenwood, Iowa. This sample came from Schuster & Knox, Schuyler, Nebr., and cost 13 cents. It was labeled: "Glenwood tomatoes. Packed by the New Glenwood Canning Co., Glenwood, Iowa. When partaking of these goods, please note the flavor. Put up fresh, solid packed, and warranted." On the label was a picture of a man holding up a pink globe on which a yellow map of Iowa took up most of the space. Above this was "Imperial Iowa, the banner State."

The can was slightly corroded. Its contents were found to have received an addition of salicylic acid.

No. 11004. Fremont tomatoes. Fremont Canning Company, Fremont, Nebr. One can of this sample was bought from O. Nelson, Schuyler, Nebr., at a cost of 15 cents, and one from A. M. Parsons, also of Schuyler, at the same price. The label was: "Fremont tomatoes, Meadow-sweet brand." The words "Meadow-sweet brand" were on a small blue paster, under which and completely hidden by it was: "Packed by the Fremont Canning Co., Fremont, Neb."

The can was slightly corroded.

No. 11006. Mariner's brand tomatoes. Bassett & Fogg, Pennsville, N. J. This sample came from Towle & Morian, Schuyler, Nebr., and cost 15 cents. The label was: "Mariner's brand tomatoes. Packed by Bassett & Fogg, at Pennsville, Salem Co., N. J."

The can was slightly corroded on the interior. The contents were found to be salicylated.

No. 11673. Abbsco brand tomatoes. J. Wallace & Son, Cambridge, Md. This sample was bought from A. A. Winfield, 215 Thirteen-and-a-half street SW., and cost 12 cents. The label was: "Abbsco brand tomatoes. Jas. Wallace & Son, Cambridge, Dorchester Co., Md. The Abbsco tomatoes are fresh from the fields near the packing house, 'cold packed' and preserved with the utmost care and cleanliness. Packed at Cambridge, Dorchester Co., Md."

The can was slightly corroded.

No. 11674. Boston market tomatoes. A. Anderson, Camden, N. J. This sample was bought of A. A. Winfield, 215 Thirteen-and-a-half street SW., and cost 15 cents. It was labeled: "Boston market tomatoes; 3-lb., full weight; guaranteed; solid packed where grown. A. Anderson, Camden, N. J."

The can was bright and clean on the inside. The contents were found to contain salicylic acid.

No. 11675. Nanticoke tomatoes. I. H. Houston, Vienna, Md. This sample came from J. J. Daly, 1367 C street NW., and cost 12 cents. It was labeled: "The finest Nanticoke tomatoes; first quality. Packed by I. H. Houston, at Vienna, Dorchester Co., Md."

Salicylic acid is present in this brand.

ANALYSES OF TOMATOES.

No. 11676. Sea-side brand tomatoes. John Schwinghammer, Egg Harbor, N. J. This sample was bought from J. J. Daly, 1367 C street SW., and cost 12 cents, The label was: "Sea-side brand, solid meat tomatoes. Packed by John Schwinghammer, at Egg Harbor City, Atlantic Co., N. J. Choice quality. cold packed."

The can was clean and not corroded on the interior.

No. 11677. Tomatoes. Popular brand tomatoes. Fait & Slagle, Baltimore. This sample was bought from S. S. Tucker, 1239 C street SW., and cost 10 cents. The label read: "Tomatoes; popular brand; extra quality. Packed by Fait & Winebrenner, Baltimore City, Baltimore Co., Md. Fait & Slagle Co., successors to Fait & Winebrenner, Baltimore."

The can was clean inside. Salicylic acid was present in the contents.

No. 11678. Albion brand tomatoes. J. E. Bull, Bel-Air, Md. This sample was bought from G. G. Cornwell & Son, 1412 Pennsylvania avenue NW., and cost 15 cents. It was labeled: "Albion brand, solid meat tomatoes. First quality. Packed by Jacob E. Bull, Bel-Air, Md."

The can was slightly corroded inside. Salicylic acid was found to be present in the contents.

Tomatoes—Weights.

Serial No.	Price per can.	Weight of full can.	Weight of can.	Solid contents.	Total contents.	Weight of dry matter.	Dry matter.	Water.
	Cents.	Grams.	Grams.	Grams.	Grams.	Grams.	Per cent.	Per cent.
10002	25	1,242	170	535	1,072	55.1	5.14	94.86
11003	13	1,143	168	577	975	66.7	6.84	93.16
11004	15	1,131	188	444	943	65.4	6.94	93.06
11006	15	1,220	186	335	1,034	58.9	5.70	94.30
11673	12	1,089	159	464	930	69.4	7.47	92.53
11674	15	1,387	184	752	1,203	80.7	6.71	93.29
11675	12	1,170	177	699	993	69.1	6.96	93.04
11676	12	1,098	170	403	928	64.4	6.94	93.06
11677	10	1,143	151	593	992	69.5	6.10	93.90
11678	15	1,144	170	720	974	52.1	5.35	94.65

Tomatoes.

Serial No.	Water.	Total dry matter.	Ether extract.	Crude fiber.	Ash.	Salt.	Corrected ash.	Nitrogen.	Albuminoids.	Carbohydrates.
	Per ct.	Per ct.	Per ct.	Per ct.	Per ct.	Per ct.	Per ct.	Per ct.	Per ct.	Per ct.
10002	94.86	5.14	.152	.448	.470	.032	.438	.167	1.044	3.026
11003	93.16	6.84	.186	.542	1.068	.464	.604	.192	1.200	3.844
11004	93.06	6.94	.248	.619	1.181	.496	.685	.193	1.206	3.686
11006	94.30	5.70	.172	.441	.559	.051	.508	.177	1.106	3.422
11673	92.53	7.47	.241	.645	.529	.056	.473	.253	1.581	4.474
11674	93.29	6.71	.215	.476	.586	.077	.509	.234	1.463	3.970
11675	93.04	6.96	.237	.560	.576	.067	.509	.244	1.525	4.062
11676	93.06	6.94	.317	.473	.591	.060	.531	.217	1.356	4.203
11677	93.90	6.10	.259	.457	.472	.060	.412	.215	1.344	3.568
11678	94.65	5.35	.227	.530	.602	.052	.550	.170	1.063	2.928

Tomatoes—Calculated to dry substance.

Serial No.	Ether extract.	Crude fiber.	Ash.	Salt.	Corrected ash.	Nitrogen.	Albuminoids.	Carbohydrates.
	Per cent.	Per cent.	Per cent.	Per cent.	Per cent.	Per cent.	Per cent.	Per cent.
10002	2.95	8.72	9.15	.63	8.52	3.25	20.31	58.87
11003	2.72	7.93	15.61	6.78	8.83	2.80	17.50	56.28
11004	3.57	8.92	17.01	7.14	9.87	2.78	17.38	53.12
11006	3.01	7.74	9.80	.89	8.91	3.10	19.30	60.07
11673	3.23	8.64	7.08	.75	6.33	3.39	21.19	59.86
11674	3.20	7.09	8.73	1.15	7.58	3.48	21.75	59.21
11675	3.40	8.05	8.27	.96	7.31	3.50	22.06	58.22
11676	4.56	6.81	8.52	.86	7.66	3.13	19.56	60.55
11677	4.25	7.49	7.74	.98	6.76	3.53	22.06	58.46
11678	4.25	9.90	11.26	.98	10.28	3.18	19.88	54.71

ASPARAGUS.

Thirteen samples of canned and bottled asparagus were examined. Salicylic acid was found in 11, and its presence in another sample was thought probable. No copper was found in any sample; nor was zinc.

No. 10777. Asparagus tips. F. H. Leggett & Co., New York. This sample was bought from J. B. Bryan & Bro., 608 Pennsylvania avenue NW., and cost 25 cents. It was labeled: "Asparagus tips. Francis H. Leggett & Co., New York. Packed at Riverside, Burlington Co., New Jersey."

The can was slightly corroded. No preservative was found. No copper, lead, or zinc was detected.

No. 10779. Oyster Bay asparagus. F. H. Leggett & Co., New York. This sample was bought from Browning & Middleton, 610 Pennsylvania avenue NW., and cost 35 cents. The label was: "Oyster Bay asparagus tips. Packed for F. H. Leggett & Co., N. Y., by the growers at Glen Head, Queens Co., New York." The words "packed for F. H. Leggett & Co." were covered by a pasted label—"Packed by Scudder & Townsend."

The can was slightly corroded. Salicylic acid was present. There was a trace of lead found, but no copper or zinc.

No. 10780. Colossal asparagus. F. H. Leggett & Co., New York. This sample was bought from J. B. Bryan & Bro., 608 Pennsylvania avenue NW., and cost 50 cents. The label was: "Colossal asparagus. Francis H. Leggett & Co., New York."

This can was much corroded. No preservative could be certainly detected. There existed no copper or zinc in this sample, but there were found 104.5 mg of lead, probably mostly solder.

No. 10781. Asperges entières. Dandicolle & Gaudin, Bordeaux. This sample came from J. B. Bryan & Bro., 608 Pennsylvania avenue NW., and cost 75 cents. The label was: "Asperges entières. Dandicolle & Gaudin, Bordeaux."

The can was much corroded. Salicylic acid was found to be present. There was no copper or zinc, but 56.4 mg of lead were found, possibly occurring as solder, of which Dandicolle & Gaudin were prodigal in this sample. Lumps of metal jutted out on all solder lines.

No. 10962. Asparagus. J. Broadmeadow & Son, Shrewsbury, N. J. This sample was bought from Robert White, jr., 900 Ninth street NW., and cost 35 cents. The label was: "Asparagus. J. Broadmeadow & Son. Packed at Shrewsbury, Monmouth Co., N. J."

The can was much corroded. Salicylic acid was found. No copper or zinc was present, but lead occurred to the extent of 9.2 mg.

No. 10963. Asparagus. Richardson & Robbins, Dover, Del. This sample was bought from John Keyworth, 318 Ninth street NW., and cost 40 cents. It was labeled: "Richardson & Robbins' asparagus. Packed at Dover, Kent County, Del., U. S. A."

The can was much corroded. Salicylic acid was found to be present. No copper or zinc was present, but there was a trace of lead.

No. 10964. Oyster Bay asparagus. Thurber, Whyland & Co., New York. This sample was bought from Elphonzo Youngs Co., 428 Ninth street NW., and cost 35 cents. The label read: "Oyster Bay asparagus tips. Thurber, Whyland & Co., New York. Packed at Moorestown, Burlington Co., N. J. All goods bearing our name guaranteed to be of superior quality and dealers are authorized to refund the purchase price in any case when consumers have cause for dissatisfaction. It is therefore to the interest of both dealers and consumers to use Thurber's brands."

The can was much corroded. A little salicylic acid was present. No lead, zinc, or copper was found.

No. 10965. Asperges en branches. R. Calbiac, San Francisco. This sample came from Charles I. Kellogg, 602 Ninth street NW., and cost 15 cents. The label read: "Asperges en branches. René Calbiac, San Francisco."

The can was slightly corroded. Some salicylic acid was present. There were 39.1 mg of lead (solder possibly) per kilo present, but no copper or zinc was found.

No. 10966. Hudson brand asparagus tips. Hudson & Co., Glen Cove, N. Y. This sample came from J. F. Page, 1210 F street NW., and cost 35 cents. The label was: "Hudson brand asparagus tips. Hudson & Co., Glen Cove, N. Y. Packed at Glen Cove, Oyster Bay, Queens Co., N. Y. All goods packed under this brand are warranted to be of the finest quality."

This can was much corroded. A very small amount of salicylic acid was found to be present. No copper or zinc existed in the sample, but there was a trace of lead.

No. 10967. Asperges. E. Du Raix, Bordeaux. This sample was bought from Geo. E. Kennedy & Co., 1209 F street NW., and cost 65 cents. The label was: "Asperges, conserves extra. Eugène Du Raix, Bordeaux."

The sample was put up in a glass jar with a lead top. Salicylic acid was present in small quantity, but there was no copper or zinc, and only a trace of lead.

No. 10968. Asperges. Dandicolle & Gaudin, Bordeaux. This sample was bought from Geo. E. Kennedy & Co., 1209 F street NW., and cost 45 cents. The label was: "Asperges. Dandicolle & Gaudin, Bordeaux."

The sample was in a glass jar with a lead top. Some salicylic acid was present, but there was no copper or zinc, and, strange to say, but a trace of lead.

No. 10969. Asparagus. R. Calbiac, San Francisco. This sample was bought from Chas. I. Kellogg, 602 Ninth street NW., and cost 15 cents. The part of the label which was legible read: *"Asparagus. René Calbiac [or 'Galbiac.'] San Francisco."*

The can was slightly corroded. Some salicylic acid was found, but no copper, lead, or zinc.

No. 11146. Pointes d'asperge. Dandicolle & Gaudin, Bordeaux. This sample was bought from J. H. Magruder, 1122 Connecticut avenue NW., and cost 40 cents a bottle. The label read: *"Pointes d'asperge. Dandicolle & Gaudin."*

Two bottles were bought of this sample. They were of glass with lead tops. As the contents differed in color they were analyzed separately. No. 11146 A was of a dark yellow color; No. 11146 B, very light. The former contained but a trace of salicylic acid, though that preservative was present in some quantity in the latter. Neither "A" nor "B" contained copper.

Asparagus—Weights.

Serial No.	Price.	Weight of full package.	Weight of package.	Weight of asparagus.	Total contents.	Dry matter.	Dry matter.	Water.
	Cents.	Grams.	Grams.	Grams.	Grams.	Grams.	Per cent.	Per cent.
10777	25	1,222	204	841	1,018	71.9	7.06	92.94
10779	35	1,570	245	805	1,325	71.6	5.40	94.60
10780	50	1,000	240	877	1,360	75.2	5.53	94.47
10781	75	1,430	242	828	1,188	58.4	4.92	95.08
10962	35	1,584	242	944	1,342	67.8	5.05	94.95
10963	40	1,080	230	585	850	50.0	5.89	94.11
10964	35	1,191	191	607	1,000	54.6	5.46	94.54
10965	15	1,095	165	660	930	42.6	4.58	95.42
10966	35	1,084	192	....	892	53.0	5.94	94.06
10967	65	2,230	1,074	986	1,150	63.3	5.48	94.52
10968	45	2,230	1,049	946	1,181	79.0	6.69	93.31
10969	15	1,145	199	700	946	48.3	5.11	94.89
11146A	40	798	375	277	423	22.0	5.19	94.81
11146B	40	795	352	242	443	27.9	6.30	92.70

Asparagus.

Serial No.	Water.	Total dry matter.	Ether extract.	Crude fiber.	Ash.	Salt.	Corrected ash.	Nitrogen.	Albuminoids.	Digestible albuminoids.	Carbohydrates.
	Per ct.	Per ct.	Per ct.	Per ct.	Per ct.	Per ct.	Per ct.	Per ct.	Per ct.	Per ct.	Per ct.
10777	92.94	7.06	.166	.657	.997	.424	.573	.386	2.413	1.355	2.827
10779	94.60	5.40	.046	.430	1.482	1.143	.339	.185	1.156	.030	2.286
10780	94.47	5.53	.053	.511	1.494	1.167	.327	.190	1.188	.706	2.284
10781	95.08	4.92	.058	.481	1.146	.814	.332	.150	.938	.636	2.297
10962	94.95	5.05	.084	.551	1.061	.690	.371	.193	1.200	.830	2.148
10963	94.11	5.89	.056	.658	1.522	1.127	.395	.210	1.350	.917	2.304
10964	94.56	5.44	.063	.525	1.099	.653	.446	.274	1.713	1.105	2.040
10965	95.42	4.58	.063	.441	.747	.378	.369	.197	1.231	.800	2.098
10966	94.06	5.94	.122	.403	1.524	1.047	.477	.324	2.025	1.612	1.776
10967	94.52	5.48	.075	.610	.861	.501	.360	.222	1.388	.705	2.516
10968	93.31	6.69	.088	.800	.935	.513	.422	.247	1.544	1.418	3.323
10969	94.89	5.11	.072	.641	.852	.475	.377	.207	1.294	1.123	2.251
11146A	94.81	5.19	.123	.560	1.209	.905	.304	.275	1.719	1.086	1.579
11146B	93.70	6.30	.080	.483	1.799	1.480	.319	.266	1.663	1.181	2.260

LABELS OF PUMPKIN SAMPLES.

Asparagus—Calculated to dry substance.

Serial No.	Ether extract.	Crude fiber.	Ash.	Salt.	Corrected ash.	Nitrogen.	Albuminoids.	Digestible albuminoids.	Per cent digestible.	Carbohydrates.
	Per ct.	Per ct.	Per ct.	Per ct.	Per ct.	Per ct.	Per ct.	Per cent.		Per ct.
10777	2.35	9.31	14.12	6.00	8.12	5.47	34.19	17.78	52.00	40.03
10779	.86	7.97	27.44	21.17	6.27	3.42	21.38	11.67	54.59	42.35
10780	.95	9.24	27.01	21.10	5.91	3.43	21.44	13.85	64.60	41.36
10781	1.18	9.77	23.29	16.55	6.74	3.04	19.00	12.93	68.05	46.76
10962	1.67	10.90	21.01	13.66	7.35	3.82	23.88	16.44	68.85	42.54
10963	.95	11.17	25.84	19.14	6.70	3.67	22.94	15.57	67.87	39.10
10964	1.16	9.65	20.21	12.00	8.21	5.04	31.50	20.31	64.48	37.48
10965	1.38	9.63	16.31	8.25	8.06	4.30	26.88	17.46	64.95	45.80
10966	2.05	8.30	25.66	17.62	8.04	5.46	34.13	27.13	79.49	29.86
10967	1.37	11.13	15.71	9.14	6.57	4.05	25.31	12.87	50.85	46.48
10968	1.32	11.96	13.93	7.66	6.32	3.69	23.06	21.19	91.89	49.68
10969	1.41	12.54	16.67	9.30	7.37	4.05	25.31	21.97	86.80	44.07
11146A	2.36	10.78	23.30	17.43	5.87	5.29	33.06	20.93	63.31	30.50
11146B	1.36	7.66	28.55	23.49	5.06	4.22	26.38	18.74	71.04	36.05

PUMPKIN.

Five samples of canned pumpkin were examined. Four contained salicylic acid.

DESCRIPTION OF SAMPLES.

No. 10782. Pumpkins. Laurel Canning Company, Laurel, Del. This sample was bought from A. A. Winfield, 215 Thirteenth-and-a-half street SW., and cost 10 cents. The label was: "Pumpkins, first quality. Packed by the Laurel Canning Co., Laurel, Del."

The can was slightly corroded inside. Salicylic acid was found.

No. 11007. Sugar pumpkin. Batavia Preserving Company, Batavia, N. Y. This sample was bought from A. M. Parsons, of Schuyler, Nebr., and cost 15 cents. It was labeled "Sugar pumpkin, unequaled in quality for old-fashioned pies. Packed by the Batavia Preserving Company, at Batavia, Genesee Co., N. Y. All goods bearing this trade-mark are guaranteed to be of the finest quality."

The can was stained a dark color on the interior. Salicylic acid was present in the contents.

No. 11679. Golden pumpkin. Curtice Brothers Company, Rochester, N. Y. This sample was bought from G. G. Cornwell & Son, 1412 Pennsylvania avenue NW., and cost 15 cents. It was labeled: "Golden pumpkin. Curtice Brothers Co., Rochester, N.Y., U.S.A. All goods under this label are of our own packing and warranted to give entire satisfaction. We guarantee the contents of this can to be of extra quality, and packed at Rochester, Monroe Co., New York, U. S. A. Our fruits and vegetables are grown in this immediate vicinity especially for our wants. Curtice Brothers Co., Preservers."

The can was slightly corroded inside. Salicylic acid was present.

No. 11682. Pumpkin. Forestville Canning Company, Forestville, N. Y. This sample was bought from J. F. Page, 1210 F street NW., and cost 12 cents. It was labeled: "Forestville brand pumpkin. Fancy quality. Forestville Canning Co., Forestville, N. Y., U. S. A. Packed at Forestville, Chautauqua Co., N. Y., U. S. A. Our guaranty embraces all products packed under the brand 'Forestville.' We will redeem the goods if not satisfactory. The Forestville Canning Co."

The can was slightly corroded inside. Salicylic acid was present in the contents.

No. 11683. Golden pumpkin. C. S. Bucklin, Keyport, N. J. This sample was bought from Elphonzo Youngs Co., 428 Ninth street NW., and cost 12 cents. The label reads: "Golden pumpkin. Packed at Keyport, Monmouth Co., N. J. C. S. Bucklin, Keyport, N. J. These goods are packed by an entirely new process, originated by us, which enables us to remove every particle of fiber; hence no sifting is required, whether used as a vegetable or for making pies. It is perfectly smooth, of a bright golden color, and retains the full flavor of the fresh vegetable. Give this brand a trial."

The can was slightly corroded. Preservatives were not found.

Pumpkin—Weights.

Serial No.	Price per can.	Weight of full can.	Weight of can.	Solid contents.	Total contents.	Weight of dry matter.	Dry matter.	Water.
	Cents.	Grams.	Grams.	Grams.	Grams.	Grams.	Per cent.	Per cent.
10782	10	1,095	180	735	915	51.8	5.66	94.34
11007	15	1,215	177	801	1,038	71.7	6.91	93.09
11679	15	1,195	172	851	1,023	61.1	5.97	94.03
11682	12	1,225	177	871	1,048	75.7	7.23	92.77
11683	12	1,235	193	849	1,042	110.7	10.62	89.38

Pumpkin.

Serial No.	Water.	Total dry matter.	Ether extract.	Crude fiber.	Ash.	Salt.	Corrected ash.	Nitrogen.	Albuminoids.	Carbohydrates.
	Per ct.	Per ct.	Per ct.	Per ct.	Per ct.	Per ct.	Per ct.	Per ct.	Per ct.	Per ct.
10782	94.34	5.66	.065	.618	.437	.036	.401	.074	.463	4.077
11007	93.09	6.91	.057	.978	.551	.015	.536	.118	.738	4.586
11679	94.03	5.97	.088	1.187	.516	.023	.493	.100	.625	3.554
11682	92.77	7.23	.083	1.109	.510	.019	.491	.080	.500	4.968
11683	89.38	10.62	.404	1.439	.520	.028	.492	.147	.919	7.338

Pumpkin—Calculated to dry substance.

Serial No.	Ether extract.	Crude fiber.	Ash.	Salt.	Corrected ash.	Nitrogen.	Albuminoids.	Carbohydrates.
	Per cent.	Per cent.	Per cent.	Per cent.	Per cent.	Per cent.	Per cent.	Per cent.
10782	1.15	10.91	7.72	.63	7.09	1.30	8.13	72.09
11007	.82	14.15	7.97	.22	7.75	1.74	10.88	66.18
11679	1.48	19.89	8.65	.38	8.27	1.68	10.50	59.46
11682	1.15	16.17	7.05	.26	6.79	1.10	6.88	68.75
11683	3.80	13.55	4.90	.26	4.64	1.38	8.63	69.12

SQUASH.

Two samples of canned squash were examined. One contained salicylic acid.

DESCRIPTION OF SAMPLES.

No. 11680. Hubbard squash. Curtice Brothers Co., Rochester, N. Y. This sample was bought from G. G. Cornwell & Son, 1412 Pennsylvania avenue NW., and cost 15 cents. It was labeled: "Hubbard squash. Extra quality. Curtice Brothers Co., Rochester, N. Y., U. S. A. All goods under this label are of our own packing, and warranted to give entire satisfaction. We guarantee the contents of this can to be of extra quality, and packed at Rochester, Monroe Co., New York, U. S. A. Our fruits and vegetables are grown in this immediate vicinity especially for our wants. This squash is carefully prepared, and will be found more economical than the fresh article."

The can was bright and clean. Salicylic acid was present in the contents.

No. 11681. Marrow squash. F. H. Leggett & Co., N. Y. This sample came from J. F. Page, 1210 F street NW., and cost 13 cents. It was labeled: "Marrow squash. Packed at Riverside, Burlington Co., N. J. Francis H. Leggett, New York."

The can was bright and clean. Preservatives were not detected.

Squash—Weights.

Serial No.	Price per can.	Weight of full can.	Weight of can.	Solid contents.	Total contents.	Weight of dry matter.	Dry matter.	Water.
	Cents.	Grams.	Grams.	Grams.	Grams.	Grams.	Per cent.	Per cent.
11680	15	1182	180	822	1002	143.9	14.37	85.63
11681	13	1277	204	869	1073	133.8	12.47	87.53

Squash.

Serial No.	Water.	Total dry matter.	Ether extract.	Crude fiber.	Ash.	Salt.	Corrected ash.	Nitrogen.	Albuminoids.	Carbohydrates.
	Per ct.	Per ct.	Per ct.	Per ct.	Per ct.	Per ct.	Per ct.	Per ct.	Per ct.	Per ct.
11680..	85.63	14.37	.060	.281	.188	.010	.178	.039	.244	3.597
11681..	87.53	12.47	.455	1.076	.615	.034	.581	.136	.850	9.474

Squash—Calculated to dry substance.

Serial No.	Ether extract.	Crude fiber.	Ash.	Salt.	Corrected ash.	Nitrogen.	Albuminoids.	Carbohydrates.
	Per cent.	Per cent.	Per cent.	Per cent.	Per cent.	Per cent.	Per cent.	Per cent.
11680..	1.37	6.43	4.31	.22	4.09	.90	5.63	82.26
11681..	3.65	8.63	4.93	.27	4.66	1.09	6.81	75.98

MACÉDOINE.

Five samples of macédoine were examined. All contained copper. Salicylic acid was contained in two, possibly in three.

DESCRIPTION OF SAMPLES.

No. 10725. Macédoine. Amieux Frères, Paris. This sample came from Frank Hume, 454 Pennsylvania avenue NW., and cost 30 cents. The label read: "Macédoine 1er choix Amieux Frères, Paris. Usine à Nantes & Perigueux."

The can was slightly corroded. No preservatives were found. Copper was present to the extent of 17.7 mg per kilo, or 7.5 mg per can. The firm omitted from this sample its usual motto, "à mieux" (for the better).

No. 10726. Macédoine. Risch & Cheminant, Paris. This sample was bought from Browning & Middleton, 610 Pennsylvania avenue NW., and cost 25 cents. The label was: "Macédoine. Risch & Cheminant, Fabrique de conserves alimentaires, Paris, France." Label (green) printed on can.

The can was bright and clean. No preservative could be identified with certainty. Copper to the extent of 27.6 mg per kilo, or 12 mg per can, was present.

No. 10727. Macédoine de légumes. Amédée Nadal, Bordeaux. This sample was bought from Browning & Middleton, 610 Pennsylvania avenue NW., and cost 25 cents. It was labeled "Macédoine de légumes. Amédée Nadal, successeur de E. Cougouille, Bordeaux. Usine à Eymet, Dordogne."

The can was bright and clean. Salicylic acid was present. Copper also existed to the extent of 35.1 mg per kilo, or 14.8 per can.

No. 10728. Macédoine de légumes; extra fine. A. Eyquem, Bordeaux. This sample was bought from J. B. Bryan & Bro., 608 Pennsylvania avenue NW., and cost 25 cents. The label was: "Macédoine de légumes; extra fine. Conserves alimentaires supérieur, préparés spécialement pour d'usage des familles. Alex^{dre} Eyquem, Bordeaux."

The can was bright and clean. No preservative was found. There were 50.4 mg of copper per kilo in this sample. This is equal to 21.4 per can.

No. 10977. Macédoine de légumes. L. A. Price, Bordeaux. This sample was bought from Geo. E. Kennedy & Co., 1209 F street NW., and cost 40 cents. The label was: "Macédoine de légumes. L. A. Price, Bordeaux, France."

The sample was in a glass jar with a tin top. No preservative was found. Copper was present to the extent of 48.4 mg per kilo, or 18.7 per bottle. There was also zinc, 3.6 mg per kilo being found.

Composition of macédoine.

Serial No.	Peas.	String beans.	Flat beans.	Turnips.	Carrots.
	Per cent.	Per cent.	Per cent.	Per cent.	Per cent.
10725	9.00	26.08	17.10	17.10	30.72
10726	13.08	24.74	8.39	10.24	43.55
10727	11.31	35.29	1.49	13.52	20.30
10728	15.56	16.25	17.07	20.30	30.82
10977	10.34	17.32	19.96	33.40	18.98

ANALYSES OF MACÉDOINE.

Macédoine—Weights.

Serial No.	Price.	Weight of full package.	Weight of package.	Solid contents.	Total contents.	Dry matter.	Dry matter.	Water.
	Cents.	Grams.	Grams.	Grams.	Grams.	Grams.	Per cent.	Per cent.
10725	30	528	102	301	426	36.2	8.50	91.50
10726	25	520	85	280	435	31.2	7.18	92.82
10727	25	509	89	236	420	17.3	4.13	95.87
10728	25	515	90	270	425	30.3	7.14	92.86
10977	40	817	430	287	387	30.0	7.77	92.23

Macédoine.

Serial No.	Water.	Total dry matter.	Ether extract.	Crude fiber.	Ash.	Salt.	Corrected ash.	Nitrogen.	Albuminoids.	Digestible albuminoids.	Carbohydrates.
	Per ct.	Per ct.	Per ct.	Per ct.	Per ct.	Per ct.	Per ct.	Per ct.	Per ct.	Per ct.	Per ct.
10725	91.50	8.50	.025	.713	1.142	.803	.339	.254	1.588	1.387	5.032
10726	92.82	7.18	.015	.728	.824	.549	.275	.215	1.344	1.171	4.269
10727	95.87	4.13	.012	.447	1.130	.925	.205	.104	.650	.522	1.891
10728	92.86	7.14	.017	.644	1.192	.877	.315	.219	1.369	1.221	3.918
10977	92.23	7.77	.020	.720	.911	.655	.256	.272	1.700	1.485	4.419

Macédoine—Calculated to dry substance.

Serial No.	Ether extract.	Crude fiber.	Ash.	Salt.	Corrected ash.	Nitrogen.	Albuminoids.	Digestible albuminoids.	Per cent digestible.	Carbohydrates.
	Per ct.	Per ct.	Per ct.	Per ct.	Per ct.	Per ct.	Per cent.	Per cent.		Per ct.
10725	.29	8.39	13.43	9.45	3.98	2.99	18.69	16.32	87.32	59.20
10726	.21	10.14	11.47	7.65	3.82	3.00	18.75	16.31	86.90	59.43
10727	.29	10.81	27.35	22.39	4.96	2.52	15.75	12.65	80.32	45.80
10728	.24	9.02	16.09	12.28	4.41	3.07	19.19	17.10	89.11	54.86
10977	.26	9.26	11.73	8.43	3.30	3.50	21.88	19.17	87.34	56.87

SUCCOTASH.

Ten samples of succotash were examined. But one contained salicylic acid. There was no copper in any of them. Zinc in small quantity was present in one sample.

DESCRIPTION OF SAMPLES.

No. 10747. Red seal succotash. T. Roberts & Co., Philadelphia. This sample was purchased from N. H. Shea, 632 Pennsylvania avenue, and cost 15 cents. It was labeled: "Red seal brand, choice quality succotash. Packed for Thos. Roberts & Co., Philadelphia, at Frederick, Frederick Co., Md. Trade-mark: The great seal of Maryland [with the inscription], Fatti maschi, parole famine. Thos. Roberts & Co. are the owners of the red seal trade-mark and so recorded in the Patent Office at Washington, D. C. Any infringement will be dealt with to the full extent of the law."

The can was badly corroded. No preservatives were found. Zinc was present in small quantities, 4.8 mg per kilo, 3.1 mg per can. There was no copper. Lead was found to the amount of 44 mg per kilo. This may have been present as solder.

No. 10748. Maryland brand succotash. T. J. Meyer & Co., Baltimore. This sample came from G. C. Burchard, 354 Pennsylvania avenue NW., and cost 13 cents. The label was: "Maryland brand, first quality succotash. Packed by T. J. Meyer & Co., Baltimore, Baltimore Co., Md."

The can was slightly corroded. No preservatives were found. No copper or zinc was found, but there was considerable lead, 47.2 mg per kilo. This was possibly due at least in part to finely divided solder.

No. 10954. Rangeley succotash. A. H. Burnham, Waterford, Me. This sample was purchased from E. E. Berry, stand 1, Riggs Market, and cost 15 cents. The label was: "Rangeley succotash. Packed by A. H. Burnham at Waterford, Oxford Co., Me. Selected and packed with special care for finest quality Maine succotash."

The can was slightly corroded. No preservatives were found. There was a trace of lead present, but no copper or zinc was found.

No. 10955. Succotash. Wayne County Preserving Company, Fairport, N. Y. This sample was bought from F. L. Bubb, stand 64, Riggs Market, at a price of 15 cents. It was labeled: "Succotash. Wayne County Preserving Co. Extra quality. Packed at Fairport, Monroe Co., N. Y., by A. H. Cobb."

The can was badly corroded. No preservatives were found. No copper or zinc was present, but a trace of lead was reported.

No. 10956. Succotash. Githens & Rexsamer, Philadelphia. This sample of succotash was purchased from John P. Love, 1534 Fourteenth street NW., for 20 cents. The label was: "Succotash. Githens & Rexsamer, Philadelphia. These goods are of unsurpassed quality."

The can was slightly corroded. No preservative was found. No copper or zinc was present, but a trace of lead occurred.

No. 10957. Popular brand succotash. Fait & Winebrenner, Baltimore. This sample was bought of Robert White, jr., 900 Ninth street NW., for 12½ cents per can. The label read: "Popular brand succotash. Fait & Winebrenner. Extra quality. Packed by Fait & Winebrenner, Baltimore City, Baltimore Co., Md."

The can was slightly corroded. No preservative was found. No zinc was present. Lead to the amount of 25.2 mg per kilo occurred, but of course may have been present as solder.

No. 10958. Onondaga cream succotash. Merrell & Soule, Syracuse, N. Y. This sample was bought from J. F. Page, 1210 F street NW., and cost 16 cents. The label was: "Onondaga cream succotash. Merrell & Soule, Syracuse, N. Y. Indian brand; extra quality. Packed at Chittenango, Madison Co., N. Y."

The can was slightly corroded. No preservative was found.

No. 10959. Succotash. Baile & Stouffer, New Windsor, Md. This sample was purchased from J. F. Russell, 730 Ninth street NW., for 15 cents. It was labeled: "Succotash; first quality; expressly for family use. Baile & Stouffer, New Windsor, Carroll Co., Md."

The can was badly corroded. In the contents both sulphurous and salicylic acids were found. No copper or zinc was found, but lead to

ANALYSES OF SUCCOTASH. 1143

the amount of 16.4 mg per kilo occurred. This, of course, may have been in the shape of finely divided solder.

No. 10960. Pride of the valley succotash. Frederick City Packing Company, Frederick, Md. This sample was bought of Elphonzo Youngs Co., 428 Ninth street, for 18 cents. The label was: "Pride of the valley brand succotash. Packed by the Frederick City Packing Co., Frederick City, Md."

The can was badly corroded. No preservative was found. No copper, lead, or zinc could be detected.

No. 10961. Paris succotash. Burnham & Morrill, Portland, Me. This sample was bought of Chas. I. Kellogg, 602 Ninth street, and cost 15 cents. It was labeled: "Paris succotash. Burnham & Morrill, Portland, Maine. U. S. A. Extra quality. Packed at Paris, Oxford Co., Maine, U. S. This succotash is a combination of Paris sugar corn and Lima beans."

The can was much corroded. No preservative was found. No copper, lead, or zinc could be detected.

Succotash—Weights.

Serial No.	Price	Weight of full can.	Weight of can.	Weigh of solid contents.	Total contents.	Dry matter.	Dry matter.	Water.
	Cents.	Grams.	Grams.	Grams.	Grams.	Grams.	Per cent.	Per cent.
10747	15	718	134	584	584	146.3	25.06	74.94
10748	13	719	121	519	598	137.2	22.95	77.05
10954	15	765	135	601	630	141.4	22.45	77.55
10955	15	776	122	654	654	167.6	25.63	74.37
10956	20	783	126	657	657	188.2	28.64	71.36
10957	12½	725	120	407	605	124.8	20.63	79.37
10958	16	752	121	631	631	159.8	25.32	74.67
10959	15	696	125	470	571	115.0	20.14	79.86
10960	18	719	124	578	595	145.9	24.52	75.84
10961	15	754	134	620	620	144.3	23.28	76.72

Succotash.

Serial No.	Water.	Total dry matter.	Ether extract.	Crude fiber.	Ash.	Salt.	Corrected ash.	Nitrogen.	Albuminoids.	Digestible albuminoids.	Carbohydrates.
	Per ct.	Per ct.	Per ct.	Per ct.	Per ct.	Per ct.	Per ct.	Per ct.	Per ct.	Per ct.	Per ct.
10747	74.94	25.06	.942	.905	.739	.170	.569	.521	3.256	2.814	19.218
10748	77.05	22.95	.730	1.010	1.448	.645	.803	.585	3.656	3.204	16.106
10954	77.55	22.45	.988	.983	.703	.020	.683	.465	2.906	2.611	16.870
10955	74.37	25.63	.720	1.138	1.146	.338	.808	.623	3.894	3.581	18.732
10956	71.36	28.64	1.103	1.103	.977	.023	.954	.662	4.138	3.725	21.319
10957	79.37	20.63	.761	.813	.856	.326	.530	.528	3.300	3.033	14.900
10958	74.68	25.32	.858	.853	.805	.329	.476	.641	4.006	3.157	18.798
10959	79.86	20.14	.775	1.124	.963	.348	.615	.548	3.425	2.904	13.853
10960	75.48	24.52	.826	.858	.915	.152	.763	.515	3.214	2.906	18.707
10961	76.72	23.28	1.001	.740	.768	.023	.745	.547	3.419	3.131	17.352

Succotash—Calculated to dry substance.

Serial No.	Ether extract.	Crude fiber.	Ash.	Salt.	Corrected ash.	Nitrogen.	Albuminoids.	Digestible albuminoids.	Per cent digestible.	Carbohydrates.
	Per ct.	*Per ct.*	*Per ct.*	*Per ct.*	*Per ct.*	*Per ct.*	*Per ct.*	*Per ct.*		*Per ct.*
10747	3.76	3.61	2.95	.68	2.27	2.08	13.00	11.23	86.39	76.68
10748	3.18	4.40	6.31	2.81	3.50	2.55	15.94	13.96	87.58	70.17
10954	4.40	4.38	3.13	.09	3.04	2.07	12.94	11.63	89.88	75.15
10955	2.81	4.44	4.47	1.32	3.15	2.43	15.19	13.97	91.97	73.09
10956	3.85	3.85	3.41	.80	2.61	2.31	14.44	13.00	90.02	74.45
10957	3.69	3.94	4.15	1.58	2.57	2.56	16.00	14.70	91.88	72.22
10958	3.30	3.37	3.18	.13	3.05	2.53	15.81	12.47	78.87	74.25
10959	3.85	5.58	4.78	1.73	3.05	2.72	17.00	14.42	84.82	68.79
10960	3.37	3.50	3.73	.62	3.11	2.10	13.13	11.85	90.25	76.27
10961	4.30	3.18	3.30	.10	3.20	2.35	14.69	13.45	91.56	74.53

MIXED CORN AND TOMATOES.

Two samples of mixed corn and tomatoes were examined. Salicylic acid was found in one and its existence in the other was probable, although it could not be identified with certainty. One sample contained a trace of zinc.

No. 10974. Corn and tomatoes. C. Foos, Baltimore. This sample was bought from F. L. Bubb, stand 64, Riggs Market, and cost 10 cents. The label was: "People's brand, first quality, corn and tomatoes. Packed by C. Foos, No. 10, Brown's Lane, Baltimore County, Md."

The can was slightly corroded. Salicylic acid was present. Zinc was also found to the extent of 3.6 mg per kilo. There was no lead or copper.

No. 10975. Corn and tomatoes. John McShane, Alberton, Md. This sample was bought from Robert White, jr., 900 Ninth street NW., and cost 10 cents. The label was: "Corn and tomatoes. Packed by John McShane on Mount Hebron Farm, Alberton, Howard County, Md."

The can was badly corroded. No preservative could be certainly identified. There was no copper or zinc and but a trace of lead.

Corn and tomatoes—Weights.

Serial No.	Price.	Weight of full package.	Weight of package.	Weight of solid contents.	Total contents.	Dry matter.	Dry matter.	Water.
	Cents.	*Grams.*	*Grams.*	*Grams.*	*Grams.*	*Grams.*	*Per cent.*	*Per cent.*
10974	10	686	124	562	562	47.7	8.48	91.52
10975	10	706	128	519	578	95.0	16.43	83.57

Corn and tomatoes.

Serial No.	Water.	Total dry matter.	Ether extract.	Crude fiber.	Ash.	Salt.	Corrected ash.	Nitrogen.	Albuminoids.	Digestible albuminoids.	Carbohydrates.
	Per ct.	*Per ct.*	*Per ct.*	*Per ct.*	*Per ct.*	*Per ct.*	*Per ct.*	*Per ct.*	*Per ct.*	*Per ct.*	*Per ct.*
10974	91.52	8.48	.338	.385	.528	.059	.469	.193	1.206	1.134	6.023
10975	83.57	16.43	.430	.598	1.100	.075	.515	.340	2.125	1.886	12.078

ANALYSES OF MIXED OKRA AND TOMATOES.

Corn and tomatoes—Calculated to dry substance.

Serial. No.	Ether extract.	Crude fiber.	Ash.	Salt.	Corrected ash.	Nitrogen.	Albuminoids.	Digestible albuminoids.	Per cent digestible.	Carbohydrates.
	Per ct.	Per ct.	Per ct.	Per ct.	Per ct.	Per ct.	Per cent.	Per cent.		Per ct.
10974	2.99	4.54	6.23	.09	5.54	2.28	14.25	13.37	93.82	70.99
10975	2.67	3.04	7.24	4.11	3.13	2.07	12.94	11.48	90.78	73.51

MIXED OKRA AND TOMATOES.

Three samples of mixed okra and tomatoes were examined. All contained salicylic acid. One also contained copper.

DESCRIPTION OF SAMPLES.

No. 10771. Okra and tomatoes. F. H. Leggett & Co., New York. This sample was bought from J. B. Bryan & Bro., 608 Pennsylvania avenue NW., and cost 20 cents. The label was: "Okra and tomatoes. Francis H. Leggett & Co., New York. Packed at Riverside, Burlington Co., N. J."

The can was bright and clean. Salicylic acid was found in the contents.

No. 10970. Okra and tomatoes. Githens & Rexsamer, Philadelphia. This sample came from John P. Love, 1534 Fourteenth street NW., and cost 15 cents. The label read: "Okra and tomatoes, Githens & Rexsamer, Philadelphia. These goods are of unsurpassed quality."

Salicylic acid was found. Copper also was found in this sample to the extent of 22.4 mg per kilo (13.2 mg per can). Lead in small amount (7.2 mg per kilo) was also present, possibly, however, in the shape of solder. Zinc was not found.

No. 10971. Okra and tomatoes. Gordon & Dilworth, New York. This sample was purchased from J. F. Russell, 730 Ninth street NW., and cost 20 cents. This label was: "Okra and tomatoes, Gordon & Dilworth, 563 and 665 Greenwich st., N. Y. A few years since okra was scarcely known except in the southern portion of the country, where it had long been used in preparing soups, etc., under the name of gumbo, for which purpose it is now extensively used and by many is esteemed as a table vegetable by itself. Combined in this manner with tomatoes its application to any kind of soup is perfect."

The can was corroded. Salicylic acid was found in the contents. No lead, copper, or zinc could be detected.

Okra and tomatoes—Weights.

Serial No.	Price.	Weight of full can.	Weight of can.	Total contents.	Dry matter.	Dry matter.	Water.
	Cents.	Grams.	Grams.	Grams.	Grams.	Per cent.	Per cent.
10771	10	1172	178	994	82.9	8.34	91.66
10970	15	734	140	594	51.0	8.59	91.41
10971	20	1526	294	1232	94.7	7.69	92.31

Okra and tomatoes.

Serial No.	Water.	Total dry matter.	Ether extract.	Crude fiber.	Ash.	Salt.	Corrected ash.	Nitrogen.	Albuminoids.	Digestible albuminoids.	Carbohydrates.
	Per ct.	Per ct.	Per ct.	Per ct.	Per ct.	Per ct.	Per ct.	Per ct.	Per ct.	Per ct.	Per ct.
10771	91.66	8.34	.200	.515	1.755	1.199	.556	.198	1.238	.912	4.563
10970	91.41	8.59	.190	.436	1.607	.965	.642	.177	1.106	.801	5.251
10971	92.31	7.69	.270	.611	1.391	.650	.741	.186	1.163	1.027	4.255

Okra and tomatoes—Calculated to dry substance.

Serial No.	Ether extract.	Crude fiber.	Ash.	Salt.	Corrected ash.	Nitrogen.	Albuminoids.	Digestible albuminoids.	Per cent digestible.	Carbohydrates.
	Per ct.	Per ct.	Per ct.	Per ct.	Per ct.	Per ct.	Per cent.	Per cent.		Per ct.
10771	3.22	6.17	21.04	14.38	6.66	2.37	14.81	10.93	73.80	54.76
10970	2.21	5.07	18.71	11.23	7.48	2.06	12.88	9.32	72.36	61.13
10971	3.51	7.95	18.09	8.45	9.64	2.42	15.13	13.35	75.02	55.32

LIST OF PACKERS WHOSE GOODS WERE EXAMINED.

In the analyses of canned goods in the preceding pages it will be noticed that, although many samples are mentioned, the packers whose goods they represented are comparatively few in number. For instance, one firm might put up a dozen kinds of vegetables, and these analyses are naturally scattered through corn, succotash, beans, etc. For this reason it was thought well to collate the results obtained from all the goods of any given packer in regard to metallic contaminations and preservatives. In the following alphabetical list will be found summarized these general conclusions in regard to all packers whose names occurred on goods analyzed in this investigation.

M. Ader & Cie., Paris.—Only one brand of this packer's goods (No. 10886—peas) was examined. It contained a large amount of salicylic acid and 31.6 mg of copper per kilo.

Amherst Packing Co., North Amherst, Ohio.—Two samples of this firm's products (Nos. 10944 and 10950), both canned beans, were examined. Both contained salicylic acid, the latter sample giving a remarkably large amount. No. 10944 contained a little zinc, but no copper, while No. 10950 contained both zinc and copper in small amounts. The amounts of these metals in both cases were too small to make it probable that they were added intentionally.

Amieux Frères, Nantes and Paris.—In one sample of peas put up by this firm (No. 10891) salicylic acid was present in relatively large amount; in another (No. 10873) it was not detected. A sample of beans (No. 10941) also contained it, while it was not found in macédoine (No. 10725). All four samples contained copper, two in excessive amounts. A pea sample (No. 10873) contained, in addition to the copper, a large amount of zinc.

A. Anderson, Camden, N. J.—But one sample of this packer's goods was examined (No. 11674—tomatoes). It contained salicylic acid.

Anderson Preserving Co., Camden, N. J.—A sample of sweet potatoes (No. 11008) put up by this firm was examined. Neither preservatives or heavy metals could be found.

.Atlantic Canning Company, Atlantic, Iowa.—One sample of corn (No. 10989) of this firm's packing was examined. It contained salicylic acid, but no copper or zinc.

Atlantic Canning Company (McWaid & Martin), Atlantic, Iowa.—One sample of corn (No. 10994) of this firm's packing was examined. It showed the presence of sulphurous and salicylic acids. No copper or zinc was found. This firm is probably the same as the preceding.

Vve. Aubin-Salles, Nantes.—One sample of peas (No. 10871) put up by this packer was examined. It contained copper and salicylic acid.

Aughinbaugh Canning Co., Baltimore.—One sample of Lima beans (No. 10995) packed by this firm was examined. It contained sulphurous acid, but no copper or zinc.

Austin, Nichols & Co., New York.—One sample of corn (No. 10911) put up by this firm was examined. It contained sulphurous and salicylic acids, and also a small amount of zinc, probably accidentally introduced.

Baile & Stouffer, New Windsor, Md.—One sample of succotash (No. 10959) and one of corn (No. 10765) of this firm's packing were examined. Both contained salicylic acid. No. 10959 also gave indications of sulphurous acid.

J. C. Baker, Aberdeen, Md.—One sample of corn (No. 10755) was examined. It contained no copper, zinc, or salicylic acid.

Baker & Brown, Aberdeen, Md.—One sample of wax beans (No. 10943) examined contained a trace of zinc and some salicylic acid. There was no copper.

Thos. W. Bamberger & Co., Baltimore.—The sample of string beans (No. 10925) examined contained salicylic acid. There was no copper.

Bamberger & Brewington, Baltimore.—Salicylic acid was found in a sample of string beans (No. 10924) put up by this firm.

Baron, Père & Fils, Paris.—One sample of peas (No. 10889) was examined. It contained no salicylic acid or other preservative. There was, however, a large amount of copper and some zinc.

Bassett & Fogg, Pennsville, N. J.—A sample of tomatoes (No. 11006) contained salicylic acid.

Batavia Preserving Company, Batavia, N. Y.—Five samples of canned goods packed by this firm were examined and the presence of salicylic acid demonstrated in four: No. 10997, Lima beans; No. 10998, red kidney beans; No. 10760, corn, and No. 11007, pumpkin. Copper was present in a sample of peas, No. 10985.

Blair Canning Co., Blair, Nebr.—A sample of corn (No. 10991) showed the presence of 20 mg of zinc per kilo. Salicylic and sulphurous acids were present.

Bonney, Wheeler, Dingley & Co., Farmington, Me.—A sample of corn (No. 10993) put up by this firm contained both salicylic acid and sulphurous acids. Copper and zinc were absent.

A. Booth Packing Co., Baltimore.—A sample of Lima beans (No. 10742) packed by this company contained sulphurous acid, but no copper or zinc.

G. H. Boyle, York Furnace, Pa.—A sample of corn (No. 10767) packed by Mr. Boyle showed a trace of zinc, but no copper or preservative. From the small amount of zinc it is probable that its presence is accidental.

Brard & Cocary, Lorient.—The title of this firm is not easy to make out from the label. It may be as given above or "Brard Cocary." One sample of peas (No. 10903) bearing this brand was examined. It contained a moderate amount of copper, but no preservative.

Charles Brewington & Co., Baltimore.—A sample of peas (No. 10697) contained no preservative or zinc. A trace of copper, possibly of accidental origin, was present.

J. Broadmeadow & Son, Shrewsbury, N. J.—A sample of asparagus (No. 10962) contained salicylic acid, but no copper or zinc.

A. F. Brown, Havre de Grace, Md.—A sample of corn (No. 10761) of this packer contained no preservative, copper, or zinc.

C. S. Bucklin, Keyport, N. J.—A sample of pumpkin (No. 11683) put up by this packer showed no preservatives.

J. E. Bull, Bel Air, Md.—A sample of tomatoes (No. 11678) put up by this packer showed the presence of salicylic acid.

A. H. Burnham, Waterford, Me.—One sample of succotash (No. 10954) contained no preservative, but a sample of corn (No. 10909) contained both sulphurous and salicylic acids. Neither sample contained copper or zinc.

Burnham & Morrill, Portland, Me.—A sample of baked beans (No. 10951) showed a large amount of salicylic acid. A succotash sample (No. 10961) showed no preservative. Neither sample contained copper or zinc.

George Cadeau & Cie.—A sample of peas (No. 10892) from this firm contained a large amount of copper. Preservatives were absent. The Massachusetts board of health (page 1160) found copper in Cadeau's peas in 1891.

René Calbiac, San Francisco.—Two samples of asparagus (Nos. 10965 and 10969) were examined. They differed from each other in style of packing. They appeared to have been in stock at the retailer's for several years. Both contained salicylic acid, but neither contained copper or zinc.

H. P. Cannon, Bridgeville, Del.—A sample of peas (No 10901) from this packer was examined. No preservative, zinc, or copper was found.

Cicero Canning Company, Chicago.—A sample of peas (No. 10984) from this firm showed no preservative, copper, or zinc.

F. Cirio, Turin, Italy.—A sample of peas (No. 10719) showed the presence of some copper, but there was no preservative found.

T. Clagett, Upper Marlboro, Md.—A sample of corn (No. 10768) from this packer contained a mere trace of zinc, and no copper or preservative.

Thos. W. Clark & Son, Glenville, Md.—A sample of peas (No. 10705) bore this name. The label stated, however, that packing was done by Fait & Winebrenner. No copper, zinc, or preservative was found.

Couteau, Paris.—A sample of peas (No. 10872) from this packer contained a large amount of copper and some zinc, but no preservative.

C. Couteaux, Paris.—A sample of peas (No. 10720) of this packer's brand which was examined contained no preservative. Copper in large amount and some zinc were present.

J. T. Cox, Bridgeton, N. J.—A sample of peas (No. 10696) from this packer contained no preservative or zinc. Copper was present as a trace.

Curtice Brothers Co., Rochester, N. Y.—Eight samples packed by this firm were examined and six found to contain salicylic acid. Its presence was probable in a seventh. Copper was present in a sample of peas (No. 10981) and a sample of baked beans (No. 10775). Zinc and sulphurous acids were not found. The eight samples may be found under peas, No. 10981; stringless beans, Nos. 10740 and 10935; baked beans, Nos. 10775 and 11001; corn, No. 10987; squash, No. 11680, and pumpkin, No. 11679.

Dandicolle & Gaudin, Bordeaux.—In two samples of peas, Nos. 10722 and 10885, and one of Brussels sprouts, No. 10979, copper in large amounts was found. Out of seven samples examined, salicylic acid was found in six, being absent from the first mentioned pea sample. In an asparagus sample, No. 11146, it was found in one bottle and not in another, or rather only in small traces. Zinc was not found in any sample. All these samples except No. 10781 were contained in glass bottles with lead tops, nothing intervening between the food and the top.

A quantity of goods packed by Dandicolle & Gaudin was seized by the authorities of Bremen on suspicion of containing copper. On analysis this was confirmed.

The French minister of commerce, on learning this, ordered the formation of a commission (Bussy and Wurtz)[1] to examine the goods put up by these packers to ascertain if they were of such nature as to prejudice French commerce. Samples were seized at the cannery in Bordeaux and at the agency in Paris. All the samples were found to have a yellowish hue, and none contained copper.

The Massachusetts board of health, in its investigations in 1889 and 1891, found copper in peas, beans, and Brussels sprouts, put up by Dandicolle & Gaudin.

---

[1] Recueil des trav. du Comité consultatif d'hygiène publique, etc., 1878, **8**, 371.

In 1887 the Brooklyn board of health forbade[1] local retailers to sell Dandicolle & Gaudin's string beans, the reason given being that these goods were heavily coppered.

Davis, Baxter & Co., Portland, Me.—A sample of corn (No. 10922) put up by this firm was examined. It was found to contain both salicylic acid and sulphurous acid.

G. W. Dunbar's Sons, New Orleans.—A sample of okra (No. 10973) and one of artichokes (No. 11217) were the only samples of this firm's goods examined. The okra was free from preservatives, copper, and zinc. The artichokes contained some salicylic acid and a little copper.

Jules Dupont, Paris.—A sample of peas, No. 10717, contained copper, but no preservative or zinc.

Eugène Du Raix, Bordeaux.—All samples of this packer's goods examined were put up in lead-topped bottles, similar to those used by Dandicolle & Gaudin. Five samples were examined. They will be found under the numbers, 10879, peas; 10936, haricots verts; 10937, haricots flageolets; 10976, haricots panachés, and 10967, asparagus. All but the first mentioned contained salicylic acid, though it was only present in a small amount in No. 10937. All the samples save the asparagus contained copper. Zinc was not found. In sample No. 10937 lead to the enormous amount of 46 mg per kilo was found; No. 10879 contained 35.2 mg per kilo, and in 10976 it existed to the amount of 15.6 mg. In No. 10967 it existed as a trace. It was undoubtedly derived from the tops. Goods packed in this manner are undoubtedly dangerous to health. This style of packing is in direct violation of the French law, which prohibits the use of alloys rich in lead in places where they may come into contact with food.

Erie Preserving Co., Buffalo, N. Y.—A sample of corn (No. 10916) contained salicylic acid to some extent, but no copper or zinc.

Evans, Day & Co., Baltimore, Md.—Two samples of Lima beans put up by this firm were examined. One (No. 10940) contained considerable salicylic acid, and in the other (No. 10745) it also appeared to be present, though it could not certainly be identified. Both contained sulphurous acid, but in neither was zinc or copper found.

Excelsior Canning Co., Maurertown, Va.—A sample of string beans (No. 10733) contained salicylic acid, but no copper.

Eyquem, Bordeaux.—A sample of macédoine (No. 10728) contained a large amount of copper. Preservatives were not found.

Fait & Slagle, Baltimore, Md.—This firm succeeded Fait & Winebrenner, according to the label of No. 11677 (tomatoes). This sample contained salicylic acid.

Fait & Winebrenner, Baltimore, Md.—Three samples of this firm's goods were examined. None contained salicylic acid or other preservatives. They were pea sample (No. 10704), okra sample (No. 10769), and

---

[1] Annual Report of Brooklyn Board of Health, 1887.

succotash sample (No. 10957). The first (No. 10704) contained copper and zinc. Another sample (No. 10705, peas) was credited both to this firm and to Thos. W. Clark & Son, Glenville, Md. It contained no preservative, copper, or zinc. No. 10704 was also credited to Nunley, Hynes & Co.

J. S. Farren & Co., Baltimore, Md.—One sample of string beans (No. 10734) examined showed no preservative, copper, or zinc.

J. Fiton, Aîné & Cie., Bordeaux.—Two brands of peas were examined. Both contained copper. They are given as Nos. 10896 and 10897. The latter gave a good test for salicylic acid. Neither contained zinc.

Fontaine, Paris.—A sample of artichokes (No. 11216) examined contained no preservative. There was a trace of copper.

Fontaine Frères, Paris.—A sample of haricots verts (No. 11214) examined contained both salicylic acid and copper.

Franklinville Canning Company, Franklinville, N. Y.—A sample of peas (No. 10700) showed the presence of a small amount of copper, but no preservative or zinc.

Frederick City Packing Co., Frederick, Md.—A sample of succotash (No. 10960) gave no evidence of the presence of preservatives, copper, or zinc.

Fremont Canning Co., Fremont, Nebr.—A sample of tomatoes (No. 11004) gave indications of the presence of salicylic acid, but this preservative could not be certainly identified.

C. Foos, Baltimore, Md.—A sample of mixed corn and tomatoes (No. 10974) was examined. It contained salicylic acid and a trifling amount of zinc. There was no copper.

Forestville Canning Company, Forestville, N. Y.—A sample of corn (No. 10918) gave evidence of having been sulphured. It contained no salicylic acid, copper, or zinc. A sample of pumpkin (No. 11682) contained salicylic acid.

W. L. Gardner, Jessups, Md.—A sample of stringless beans (No. 10931) was examined. It contained salicylic acid and a mere trace of copper, but no zinc.

Vve. Garres, jne., & Cie., Bordeaux.—Two samples of peas were examined, Nos. 10629 and 10887. The former contained a large amount of zinc and the latter almost as large an amount of copper. Neither contained preservatives.

Gibbs Preserving Co., Baltimore, Md.—A sample of peas (No. 10708) which was examined contained large amounts of both salicylic acid and copper. No zinc was present.

Githens & Rexsamer, Philadelphia, Pa.—Five samples were examined. Four contained salicylic acid. They were No. 10913 (corn), which also contained sulphurous acid; No. 10934 (stringless beans); No. 10926 (string beans), and No. 10970 (mixed okra and tomatoes), which last also contained copper. No. 10956 (succotash) was free from preservatives or copper. None of these samples contained zinc.

Githens, Rexsamer & Co., Philadelphia, Pa.—Two samples of peas (Nos. 10898 and 10899) and one of okra (No. 10972) were examined. None showed the presence of preservatives or of zinc. Nos. 10898 and 10972 contained copper. No. 10899 showed a mere trace of copper.

Glenwood Canning Co., Glenwood, Iowa.—A sample of tomatoes (No. 11003) which was examined contained salicyclic acid.

Gobelin, Fils & Cie., Paris.—Two pea samples were examined, Nos. 10723 and 10875. Both contained copper, and the latter, in addition, a very large amount of zinc. Neither showed evidence of the presence of chemical preservatives.

A. Godillot, Bordeaux.—Two samples of green beans (Nos. 10938 and 10939) were examined. Both showed the presence of salicylic acid in some quantity, and No. 10939 also that of copper. Copper was found in peas, put up by this firm, by the Massachusetts board of health in 1891 (see page 1160).

Gordon & Dilworth, New York, N. Y.—A sample of mixed okra and tomatoes (No. 10971) showed the presence of salicylic acid, but was free from copper and zinc.

Grocers' Packing Co. (Potter & Wrightington), Boston, Mass.—Two samples of baked beans were examined, Nos. 10953 and 10774. Both contained salicylic acid and a small amount of copper. No. 10774 also contained sulphurous acid. Neither contained zinc. (See "Potter & Wrightington, Boston.")

Guillaumez (Crown Imperial) Nancy.—Two samples of peas were examined. They are given as Nos. 10894 and 10895. Both contained copper in large amounts but no zinc was found. In the last-mentioned sample salicylic acid was present in some quantity. The other gave no indication of preservatives.

Hamburgh Canning Company, Hamburgh, N. Y.—This firm was represented by two samples, Nos. 10625 and 10626, both peas. The former contained a large amount of copper and the latter a much smaller quantity. Neither showed the presence of a preservative or of zinc.

C. K. Harrison, Upperville, Va.—A sample of corn from this packer was examined and is recorded as No. 10921. It contained both salicylic and sulphurous acids, but was free from copper and zinc.

H. F. Hemingway & Co., Baltimore, Md.—A sample of peas put up by this packer will be found under No. 10983. It was free from copper, zinc, and preservatives.

I. H. Houston, Vienna, Md.—Two cans of a sample of corn (No. 10749) each showed the presence of a large amount of zinc. There was no copper or preservative. A sample of tomato, recorded as No. 11675, contained salicylic acid.

Hudson & Co., Glen Cove, N. Y.—A sample of asparagus from this packer, which bears the number 10966, was examined. It contained salicylic acid but no copper or zinc.

LIST OF PACKERS.

G. W. Hunt & Co., Baltimore, Md.—A sample of peas (No. 10712) packed by this firm contained both copper and zinc, the latter in small amount. No preservative was found.

W. L. James, Hagerstown, Md.—A sample of corn (No. 10919) from this packer was examined. A trace of zinc was found. Copper was not present, and salicylic acid could not be certainly identified.

Keagle & Guider, Baltimore, Md.—A can of peas prepared by this firm was examined and figures as No. 10900. It was free from preservative, copper, and zinc.

Le Lagadec, Lorient, France.—A sample of peas, No. 10878, was examined. It was free from preservatives and zinc, but showed an excessively large amount of copper.

Charles Laing & Co., Baltimore, Md.—A sample of peas packed by this firm was examined and is recorded as No. 10884. It contained an extraordinarily large amount of salicylic acid. Copper and zinc were absent.

Henri Lambert & Cie, Bordeaux.—Two samples of peas were examined and are recorded as Nos. 10880 and 10904. Both contained large quantities of copper. Neither contained zinc or preservatives. Copper was found in this firm's peas by the Massachusetts Board of Health in 1891 (see page 1160).

H. S. Lanfair & Co., Baltimore.—A sample of string beans, No. 10736, put up by this firm contained a small amount of copper but no preservatives.

Laurel Canning Co., Laurel, Del.—A sample of pumpkin, No. 10782, put up by this firm contained salicylic acid.

Francis H. Leggett & Co., New York.—Ten samples of this firm's goods were examined. Five contained salicylic acid, the list being: Pea sample, No. 10905; corn, 10752; okra, 10770; mixed okra and tomatoes, 10771, and asparagus, 10779. Two more, an asparagus sample, No. 10780, and Lima beans, 10744, probably contained it, but the amount present was not large enough to allow it to be identified with certainty. In squash sample, No. 11681, it may also have been present. In asparagus sample, No. 10777, and pea sample, No. 10628, it was probably absent. The last-named sample contained a trace of copper, but this metal was absent from the others. A trace of zinc was present in the Lima beans.

Léopold, Bordeaux.—A sample of peas from this firm is given as No. 10874. It contained copper. No salicylic acid or other preservative was found. The Massachusetts board of health found copper in peas bearing this brand in its investigation of 1891 (see page 1160).

C. Lewis & Co., Boston, Mass.—A sample of baked beans from this firm is given as No. 10773. It contained salicylic acid. There was no copper or zinc.

Los Angeles Packing Co.—A sample of Lima beans from this firm was bought in Nebraska and is given under No. 10948. Preservatives could not certainly be identified.

Louit Frères & Co., Bordeaux.—A sample of green beans (No. 10940) from this firm showed the presence of a large amount of salicylic acid. There was no copper.

J. F. Lowekamp, Baltimore, Md.—A sample of peas from this packer, No. 10695, contained no preservatives but showed a trace of copper. In a sample of string beans, No. 10732, salicylic acid was found but there was no copper.

J. Ludington & Co., Baltimore., Md.—Three samples of peas, Nos. 10713, 10714, and 10883, were examined. All were free from zinc and Nos. 10713 and 10883 from preservatives or copper. No. 10714 contained a little copper and a large dose of salicylic acid.

Maine State Packing Co., Portland, Me.—Two brands of this firm's goods were examined, a sample of Lima beans, No. 10947, and a sample of corn, No. 10912. The corn contained salicylic acid. Neither contained zinc or copper.

E. B. Mallory & Co., Baltimore, Md.—A sample of string beans, No. 10923, from this firm contained salicylic acid. There was no copper.

Markell Brothers, Baltimore, Md.—A sample of peas from this firm, No. 10902, was examined. It showed no preservative and no zinc. Copper was present.

C. P. Mattocks, Portland, Me.—A sample of Lima beans, No. 10741, and a sample of corn, No. 10757, were examined. The former contained sulphurous acid, and the latter salicylic acid in rather large quantity. Both were free from copper and zinc.

C. A. McGaw, Perryman, Md.—A sample of corn, No. 10754, from this packer contained a large amount of salicylic acid. There was no copper or zinc.

H. J. McGrath & Co., Baltimore, Md.—Two samples of Champion brand string beans (Nos. 10927 and 10737) put up by this packer were examined. Both showed the presence of salicylic acid and 10737 contained copper.

Louis McMurray Packing Co., Frederick, Md.—One sample (No. 10746) from this packer was examined. It contained sulphurous acid and a little zinc. Salicylic acid and copper were absent.

John McShane, Alberton, Md.—A sample of mixed corn and tomatoes No. 10975, put up by this packer was examined. No preservative could be detected with certainty. Copper and zinc were absent.

Merrell & Soule, Syracuse, N. Y.—Two samples of this firm's "Onondaga" brand canned foods were examined. One was succotash, No. 10958, and the other Lima beans, No. 10945. Neither contained preservatives, copper, or zinc.

T. J. Meyer & Co., Baltimore, Md.—This may be the same firm as Thos. J. Myer. One sample of succotash, No. 10748, was examined. Preservatives, copper, and zinc were absent.

Miller Brothers & Co., Baltimore, Md.—A sample of peas, No. 10627, examined contained a large amount of salicylic acid but no zinc or copper.

Mitchell Brothers—A sample of corn from this firm (No. 10010) was examined. It contained a little zinc. There was no preservative or copper.

John Moir & Son, London.—But one sample (No. 10942) of this firm's goods was examined. This consisted of two bottles of "haricots flageolets," and was probably the best packed sample encountered, inasmuch as the vegetables did not come into contact with anything but cork and glass. Nevertheless lead was present in large quantity. Copper also existed in a small amount. No preservative could be identified with certainty.

Mound City Preserving Co., St. Louis, Mo.—Two samples bearing this brand were examined. A sample of peas (No. 10980) showed no preservatives, but contained a small amount of copper. A sample of stringless beans (No. 11000) contained both sulphurous and salicylic acids. Copper was absent.

Thos. J. Myer & Co., Baltimore, Md.—Three samples bearing this brand were examined. A pea sample (No. 10707) was free from preservatives, but contained small quantities of zinc and copper. No. 10920, corn, contained salicylic acid, but neither of the metals. No. 10729, string beans, contained a large quantity of salicylic acid and a little zinc, but no copper.

Amédée Nadal, Bordeaux.—Four samples bearing this brand were examined. They are recorded as pea samples Nos. 10715, 10716, and 10721. The peas were labeled "au naturel," and No. 10727, macédoine. All contained copper in large amounts, and 10716, in addition, a large, amount of zinc. The macédoine and No. 10715 contained salicylic acid; the latter in rather large quantity. The others were free from preservatives.

Nail City Packing Co., Wheeling, W. Va.—One sample of peas (No. 10906) was examined. It contained no copper or preservative, but zinc was present.

Nebraska City Canning Co., Nebraska City, Nebr.—One sample of corn (No. 10990) from this firm was examined. It contained some salicylic acid and sulphurous acid. Copper and zinc were absent.

S. Nicolas & Cie., Bordeaux.—One can of peas bearing this brand was examined. The results are recorded under No. 10893. Copper was present in large amount, but zinc and preservatives were absent.

Northern Maine Packing Co., Dexter, Me.—One sample of corn (No. 10917) bearing this firm's name was examined. It was free from salicylic and sulphurous acids, and also from copper and zinc.

J. Nouvialle & Cie., Bordeaux.—Two samples were examined. One of artichokes (No. 11215) contained no preservative, copper, or zinc. The other sample was No. 10890 (peas). It contained a large amount of copper. In 1891 the Massachusetts Board of Health reported copper in the peas of a firm—"J. Nouville & Cie" (see page 1160)—which was very likely the same firm.

Wm. Numsen & Sons, Baltimore, Md.—Four samples of canned goods put up by this firm were examined. Two samples of peas (Nos. 10659 and 10706) both contained copper. The former contained salicylic acid in some quantity, but the latter had none. One sample of Lima beans (No. 10743) contained both salicylic and sulphurous acids. There was no copper or zinc. A sample of string beans (No. 10660) contained no preservatives.

Nunley, Hynes & Co., Baltimore, Md.—A sample of corn (No. 10759) bearing this brand was examined. It contained salicylic acid and zinc, both in some quantity. A sample of peas (No. 10704) bears the name of this firm and also "packed by Fait & Winebrenner." Copper and zinc were present, but there was no preservative.

Parson Brothers, Aberdeen, Md.—A pea sample from this firm contained salicylic acid, copper, and zinc. It is recorded under No. 10694. This was the only sample found bearing date of packing.

C. H. Pearson Packing Co., Baltimore, Md.—One sample of string beans (No. 10999) put up by this firm was examined. It contained a very large amount of salicylic acid. Zinc and copper were absent.

Potter & Wrightington, Boston, Mass.—This firm is apparently related to the Grocers' Packing Company, also of Boston, in some way. Three of its samples of baked beans were examined. Two of these (Nos. 10006 and 10772) contained salicylic acid, and it is probable that this substance was also contained in the third (No. 10776), though it could not be identified with certainty. One sample (No. 10772) also contained sulphurous acid.

L. A. Price, Bordeaux.—Three samples were examined. Two were peas (Nos. 10877 and 10907) and one was macédoine (No. 10977). All contained large amounts of copper, and the macédoine also contained a little zinc. Preservatives were not found.

Richardson & Robbins, Dover, Del.—A sample of asparagus (No. 10963) put up by this firm was examined. It contained salicylic acid. Copper and zinc were absent.

Risch & Cheminant, Paris.—A sample of macédoine (No. 10726) from this firm was examined. It contained copper, but no zinc. A preservative could not be identified with certainty.

T. Roberts & Co., Philadelphia, Pa.—A sample of succotash (No. 10747) was examined. It contained a little zinc, but no copper or preservative.

Robinet & Cie., Nantes.—A sample of canned peas (No. 10876) was examined. It contained copper. There were no preservatives or zinc.

A. B. Roe, Greensborough, Md.—A sample of peas (No. 10701) was examined. It contained copper, but preservatives and zinc were absent.

John Root & Son, Mechanicstown, Md.—Two samples of "Pen-Mar" corn (Nos. 10751 and 10915) were examined. No. 10915 contained sulphurous acid, but salicylic acid, copper, and zinc were absent from both.

Royal Preserving Co., Chicago.—A sample of baked beans (No. 11002) was examined. It contained salicylic acid. Copper and zinc were absent.

LIST OF PACKERS.

W. C. Satterfield, Greensborough, Md.—A sample of peas (No. 10881) from this packer was examined. It contained large quantities of both copper and zinc. There was no preservative.

John Schwinghammer, Egg Harbor, N. J.—One sample of tomatoes (No. 11676) was examined. Preservatives could not be certainly identified.

B. F. Shriver, Union Mills, Md.—Nine samples of this packer's goods were examined, the list of those tested being Nos. 10699, 10702, and 10703, all peas; Nos. 10735 and 10928, string beans; No. 10932, stringless beans; and corn samples Nos. 10750, 10758, and 10766. Nos. 10699 and 10766, the first and last mentioned, were free from salicylic acid; the remaining seven contained that substance. Copper was present in the first and third pea samples. Zinc and sulphurous acid were absent from all.

A. W. Sisk, Preston, Md.—A pea sample (No. 10888) from this packer showed the presence of copper. Zinc and preservatives were absent.

Smith, Yingling & Co., Westminster, Md.—A sample of corn from this firm(No. 10910) contained some zinc, but no copper. No preservative could be detected.

Steele Bros., New Britain, Conn.—Those samples of this firm's goods which were examined, No. 10929, string beans, and No. 10930, wax beans, were put up in glass bottles with a glass top, the joint being made on a rubber ring. They presented a very neat appearance. Two bottles were bought of each sample. On examination, one rubber ring from each sample was found to contain lead sulphate, the amounts being, respectively, 1.62 and 7.54 per cent. The other ring in each case was free from lead, though it contained zinc. The samples themselves both contained lead, probably derived from the rings. In the case of 10929 the amount reached 5.2 mg per kilo, and in the other, 34.4 mg, equal in the latter case to 24.8 mg per bottle. Both samples contained salicylic acid. No. 10929 contained a small amount of copper, and No. 10930 a little zinc.

W. L. Stevens, Cedarville, N. J.—A tomato sample, No. 10002, from this packer was found to contain salicylic acid.

Charles G. Summers & Co., Baltimore, Md.—A corn sample, No. 10764, which was examined contained salicylic acid and a trifling amount of zinc. Copper was absent.

Talbot Frères, Bordeaux.—Two samples were examined, one of peas, No. 10661, and one of green beans, No. 10739. The former contained an enormous amount of copper. The beans were free from this metal, but contained salicylic acid.

Thurber, Whyland & Co., New York.—Four samples from this firm were examined, one of stringless beans, No. 10933; one of Lima beans, No. 10949; one of asparagus, No. 10964; and one of corn, No. 10914. The first three contained salicylic acid. The corn was free from this substance, but contained sulphurous acid. Copper and zinc were absent from all.

Tisserand et Fils, Paris.—A large amount of copper was found in a sample of peas, No. 10724, from this firm. The Massachusetts board of health found copper in peas from this packer in its investigations in 1891.

G. Triat & Co., Bordeaux.—A sample of peas examined, No. 10718, contained copper, but no preservative.

Van Camp Packing Co., Indianapolis, Ind.—A sample of peas, No. 10709, put up by this firm was found to contain zinc. No preservative or copper was present.

Martin Wagner Co., Baltimore, Md.—A sample of peas, No. 10698, from this packer contained copper and zinc, but no salicylic or sulphurous acid.

Jas. Wallace & Son, Cambridge, Md.—A sample of peas, No. 10882, contained both zinc and copper. A preservative was not found, nor was any found in No. 11673, a sample of tomatoes.

Wayne County Preserving Co., Newark, N. J.—A sample of baked beans, No. 10952, and one of succotash, No. 10955, were examined. The succotash was packed at Fairport, N. Y. Copper, zinc, and preservatives were absent.

Western New York Preserving & M'f'g Co., Springville, N. Y.—Two samples from this company were examined, No. 10710, peas, and No. 10762, corn. In the former there was some copper and an enormous amount of zinc. In the latter neither metal was found. Preservatives were absent from both.

W. S. Whiteford, Delta, Pa.—A sample of peas, No. 10711, contained both copper and zinc. Preservatives were absent.

R. Williamson & Co., Baltimore, Md.—A sample of corn, No. 10753, contained a small amount of zinc.

P. F. & D. E. Winebrenner, Baltimore, Md.—Two samples of this firm's goods were examined, Nos. 10008 and 10730, both string beans. In both salicylic acid was present, though it was only in small amount in the latter. Copper and zinc were absent.

J. A. Wright & Bro., Choptank, Md.—In a sample of string beans, No. 10731, preservatives could not be detected. Copper was absent.

APPENDIX.

PROHIBITION OF SALE OF COPPERED PICKLES IN BROOKLYN.[1]

DEPARTMENT OF HEALTH, OFFICE OF THE COMMISSIONER,
Brooklyn, N. Y., July 16, 1885.

To all dealers in pickles:

You are hereby notified of the following official action taken with regard to the coloring of pickles with copper:

By virtue of the power conferred upon me by law, I hereby declare the practice of coloring pickles with copper in any form to be dangerous and detrimental to public health, and do hereby prohibit the selling, or having for sale, in Brooklyn, of pickles so colored.

The above action was taken June 26, 1885.

J. H. RAYMOND,
Commissioner of Health.

SALE OF CANNED VEGETABLES COLORED WITH THE SALTS OF COPPER.[2]

At the meeting of May 5, 1891 [of the Massachusetts State board of health], the fact was presented that peas, beans, and other vegetables preserved in tin cans and glass jars, and colored with blue vitriol or other salts of copper, are sold in Massachusetts in large quantities. It appears that the manufacture of preserved foods colored by copper is confined to France. The object of this practice is evidently to give to the articles thus sold a fictitious value by making the canned vegetables appear to the consumer like those which have been freshly gathered, and the demand for them would quickly cease if they should be labeled "canned peas colored with sulphate of copper."

The practice of putting poisonous substances into food in any quantity whatever is an objectionable one, and the board therefore expressed its opinion upon the subject as follows, and has issued a circular to the same effect:

In the opinion of the board, the sale of articles of food containing such well-known poisonous substances as the salts of copper is a violation of the statutes relating to the inspection of food and drugs.

The provision of the statutes specially referred to are the following:

Chapter 263 of the Acts of 1882, section 3:

"An article shall be deemed to be adulterated within the meaning of this act—in the case of food—(6) if it is colored, coated, polished, or powdered, whereby damage is concealed, or if it is made to appear better or of greater value than it really is; (7) if it contains any added poisonous ingredient or any ingredient which may render it injurious to the health of a person consuming it."

In conformity with the foregoing action, the board has had samples of such articles examined. Canned vegetables put up by the following firms and having the following brands have been found to contain the salts of copper as an adulterant. The pub-

---

[1] Annual Report of Brooklyn Board of Health, 1885.
[2] Massachusetts Monthly Bulletin of Food and Drug Inspection, April, 1891.

lication of this list should not be taken as evidence that other brands are exempt from the same practice, since there may be many brands which have not yet come to the notice of the board:

Packer.	Brand.	Locality.
Barton Fils	Petits pois	Paris.
Bernard, Alexandre		Bordeaux.
Billet	Peas and beans	Paris.
Bouvais		Flou [?].
Brennard-Frères	Extra fino	
Briant		
Cadeau, Geo. & Cie	Pois extra fins	Paris.
Charpentier	Petits pois fins	Usine de Montrouge.
Constant Fils	Petits pois	Paris.
Couteau	Petits pois fins	Do.
Crown Imperial		Nancy.
Dadelzen, E. M		Bordeaux.
Dandicolle & Gaudin	Peas, beans, and Brussels sprouts	Do.
Dufour, E., & Cie		
Dupin, A		
Dupont, Jules		Paris.
Duprat, Clemont & Maurel		Bordeaux.
Du Raix, Eugène		Bordeaux.
Durand, Pierre		
Eyquem, Alexandre	Petits pois verts	
Fernand, Eugène, & Cie		Bordeaux.
Fiton & Cie	Petits pois verts au naturel	Do.
Fontaine Frères	Petits pois extra fins	
Vve Garres & Fils		
Gillard Fils		Bordeaux.
Godillot, Alexis, Jne		Do.
Gouleau		
Guillaumez		Nancy.
Julien, Chs		
Lambert, Henri		Bordeaux.
Laman, François & Co	Petits pois verts	Do.
Lemouies		
Leopold, Victor		
Marcelino	Petits pois fins	Paris.
Marcies & Cie		
Mercier, Eugene		Paris.
Monbadou, R		Paris and Bordeaux.
Nicholas & Cie		
Nouville, J., & Cie		
Pinard, Alphonse	Petits pois verts	Bordeaux.
Rilhac		
Rodel & Fils Frères	Petits pois extra fins au naturel	
Rolland & Grasset		Paris and Bordeaux.
Rondenet, F		Nantes.
Soule & Price		Bordeaux.
Talbot Frères		Do.
Talbot, G	Peas and beans	Do.
Tertrais, Victor	Pois moyens	Nantes.
Tisserand & Fils	Extra fine	Paris.
Triat, Gabriel, & Cie		Bordeaux.

LIST OF PACKERS USING COPPER. 1161

[ADDITION OF SULPHATE OF COPPER TO] CANNED VEGETABLES.[1]

The addition of copper sulphate in small quantities to give a green color seems to be a common practice with the following firms:

Packer.	Brand.	Locality.
Barton Fils	Peas	Paris.
Bernard, Alex	do	Bordeaux.
Billet, A	Beans	
Charpentier	Peas	Usine de Montrouge.
Dadelzen, E. M	do	Bordeaux.
Dandicolle & Gaudin	Peas, beans, sprouts	
Duprat, Clement & Maurel	Peas	
Eyquem, Alex	do	Bordeaux.
Fiton, J., ainé, & Cie	do	Do.
Fontaine Frères	do	
Guillaumez	do	Nancy.
Lanan, François, & Cie	do	Bordeaux.
Marcelino	do	Paris.
Pinard, Alphonse	do	Bordeaux.
Eugene du Raix	do	Do.
Rödel & Fils, Frères	do	
Roudenet	do	Nantes.
Soule & Price	do	Bordeaux.
Talbot, G	Beans	Do.
Tertrais, Victor	Peas	Nantes.
Triat, Gabriel, & Cie	do	Bordeaux.

IMPORTED CANNED FOODS.[2]

The following table will show the number and condition of imported canned goods examined during the past year:

	Total.	Standard.	Adulterated or not standard.
Beans	4	2	2
Capers	1	1	0
Mushrooms	4	4	0
Peas	98	12	86
	107	19	88

The chief adulterant found in these articles was copper, which was added to give a green color to the vegetables. The following names were on the labels of the adulterated samples:

Luneville, Guillaumez.
Barton Fils.
F. Roudenet & Cie.
Dandicolle & Gaudin.
Petits Pois (no packer).
R. Monbadon.
S. Nicholas & Cie.
G. Talbot.
E. Dufour & Cie.
Gillard Fils.
Duprat, Clement & Mauriel.
J. Fiton Aine & Cie.

A. Dufour & Cie.
L. A. Price.
Briaut.
A. Dupin.
Alex. Eygueme.
Gabriel Triat & Co.
Vie Garres & Fils.
Talbot Freres.
Eugene Du Raix.
Eugene Mercier.
Alphouse Pinard.
Goulean.

Chs. Julien.
A. Marci & Cie.
Bouvais, Flou.
Henri Lambert.
Rodel & Fils.
Rilhac.
Henri Lambert & Cie.
J Nouville.
Lemonie.
Victor Leopold.
Pierc Durand.

[1] Rep. Mass. State Board of Health, 1889, 86.
[2] Rep. Dairy Comm., New Jersey, 1889, 33.

APPARATUS FOR COOKING VEGETABLES AND FRUITS.[1]

A copper cooking apparatus is connected with a continuous-current dynamo in such a way that the cooking kettle is connected with the negative wire while the positive wire goes to an electrode hanging in the fluid in the kettle.

To cook fruit and vegetables in this apparatus, pass through the fluid contained in the kettle (salt water) a current of 5 ampères for about one and one-half minutes and then quickly throw in the fruit. The fruit is then to be boiled for a short time (about three minutes) the current being kept running. This ends the operation.

The fruit is now ready for Pasteurizing. It is packed in cans which are closed and soldered and placed in boiling water for half an hour.

Claim.—A copper cooking apparatus, which is connected with a continuous-current dynamo in such a way that one wire (the negative) is connected directly with the apparatus while the other (positive) is connected with an electrode hanging in the cooking fluid; and which serves for cooking fruits and vegetables, intended for preserving, so that they retain their natural color.

COPPER SULPHATE IN GREEN PEAS.[2]

James Brown, Glasgow, was arrested for having sold a can of green peas containing, according to analysis, 0.0045 per cent copper sulphate. At the trial, Dr. Sympson, professor of legal medicine at the University of Glasgow, declared copper to be a cumulative poison. Brown was fined £4.

PRESENCE OF METALLIC COMPOUNDS IN ALIMENTARY SUBSTANCES.[3]

The quantity of $CuSO_4$, usually added to preserved peas, varies between 1 and 2 grains of the ordinary blue sulphate to the can containing from $9\frac{1}{4}$ to $9\frac{3}{4}$ ounces of peas and 150 cc of liquor.

The method used for the detection of copper was to burn the sample with a mixture of sodium carbonate and potassium nitrate, dissolve in acid, and add excess of ammonia and filter from alumina, phosphates, etc.

As a result of some experiments the authors came to the conclusion that peas in stomachic digestion give up their copper to solution, but that a part, probably the greater part, passed out with the feces. Therefore, inasmuch as one person consumes only about 2 ounces of preserved peas at a meal, and this quantity contains only a fraction of a grain of $CuSO_4$, and as only a fraction of this amount is absorbed into the system, it is impossible to defend the opinion of the prejudicial influence of such amounts of copper on health. It is believed that preserved peas so colored are absolutely harmless to health.

NOTE ON COPPER IN VEGETABLES.[4]

An examination was made of a can of macédoines.

	Cu.
Peas and green vegetables contained	0.100
Carrots and turnips	.010
Water	.005
Mixed disintegrated matter	.015
Total per can	.130

Some cans of champignons and some crystallized fruits were also examined, but no copper was found in anything but some highly-colored green gages.

---

[1] J. Clot & Cie, Strassburger Konservenfabrik, Strassburg, i. E., Swiss patent No. 2019, March 25, 1890; abs. Ber. bayr. Vertr. angew. Chem. 1891, **10**, 81.

[2] Brit. and Colonial Drugg.; abs. Ztschr. Nahr. Hyg., 1891, **5**, 216.

[3] B. H. Paul and C. T. Kingzett, Analyst, 1878, **2**, 98.

[4] J. Muter, Analyst, 1878, **2**, 4.

HEAVY METALS IN FOOD.

COPPER IN PRESERVED GREEN PEAS.[1]

The method employed for copper was: Weigh 1,000 grains of peas and the liquor with which they were mixed into a porcelain basin, dry and ignite over the flame of a Bunsen burner. When the mass has burned down to a gray ash, it is cooled, moistened with H_2SO_4, and reignited. The treatment with H_2SO_4 prevents the loss of copper which would occur from the presence of NaCl in the ash at the temperature necessary to burn off the last portion of carbon. Unless carbon be wholly removed from the ash copper can not be completely dissolved from it. The ash is then boiled with acid and the copper electrolytically deposited from the solution.

The presence of copper may be beautifully shown by placing a quantity of the peas themselves in a platinum dish, acidifying with HCl, and electrolysing. In about twenty-four hours an abundant deposit of copper is obtained, but the whole of it can not be thus separated.

ANALYSIS OF CANNED PEAS.[2]

The juice around the peas contained 1.11 per cent of dry matter. The total water was 58.3 per cent, so that, as ripe peas contain about 14 per cent, about 45 per cent had been added for cooking purposes. The albuminoids calculated in the sample with 14 per cent of water were 21 per cent.

RUBBER RINGS IN THE PRESERVING INDUSTRY.[3]

Cans for preserving purposes are now made which are closed by tops made tight by rubber rings, and which have another ring pressed tightly against the solder seam at the bottom. The German law prohibits the use of solder containing more than 10 per cent of lead where it can come in contact with food, but as lead-poor solder is difficult to use, the manufacturers have adopted the rubber ring system. Foods thus preserved, however, show astonishingly high amounts of lead. This was finally explained by the analysis of the rubber rings, which showed 60 to 66 per cent of minium (red lead).

METAL VESSELS FOR CULINARY PURPOSES.[4]

The results of a series of investigations led the royal health office (Prussia) to conclude:

(1) That vessels of tinware without exception give up more or less lead to fluid contents.

(2) That upon the whole this vulnerability diminishes with increasing lead content of the alloy.

(3) That the amount given up depends on the nature of the attacking fluid, upon the mechanical condition of the metal surfaces, as well as of exterior influences (temperature and air), and also upon the kind of use made of the vessels.

(4) That the amount given up by the contents does not increase proportionally with the time of action as the dissolved lead is again precipitated partially by the tin in the course of time, and as some of the lead is also precipitated in the form of insoluble combinations with the components of the fluid.

(5) That besides vinegar other substances used for food purposes (wine, beer, solutions of salt and sugar, milk, tea, etc.) can take up lead from tinning alloys. Against all expectation in these experiments weak vinegar attacked the vessels more than stronger.

---

[1] Charles H. Piesse, Analyst, 1878, 2, 27.

[2] G. W. Wigner, Analyst, 1880, 15, 102.

[3] W. Reuss, Chem. Ztg., 1891, 15, 1522; abs, Chem. Centrbl., 1892, 1, 63.

[4] Gustav Wolffhügel, Arbeiten aus der Kais. Gesundheits Amt. 1887, 2, 112 (Berlin); abs. Chem. Centrbl., 1887, 592.

ON THE OCCURRENCE OF TIN IN ARTICLES OF FOOD AND DRINK, AND ON THE PHYSIOLOGICAL ACTION OF TIN COMPOUNDS.[1]

[Abstract.]

The method of analysis adopted was as follows: About 30 grams of the article to be examined were incinerated in a platinum dish and the ash heated with strong hydrochloric acid, the acid for the most part boiled off, about 30 or 40 cc. of water added, boiled and filtered. This alternate treatment with water and acid was repeated until sulphuretted hydrogen no longer indicated the presence of tin. The clear and, as a rule, colorless solution thus obtained was precipitated by the hydrogen sulphid in the usual manner. The following vegetable foods all gave abundant yellow precipitate of stannic sulphid: French asparagus, American asparagus, peas, tomatoes, peaches (3 brands), pineapples (2 brands), white cherries, red cherries, and marmalade. In several cases the inner surface of the canister was found much corroded. So considerable is the proportion of tin in most of the acid fruits that tin reactions can be obtained from 2 or 3 grams of the substances. A metallic taste is sometimes perceptible. The following animal foods were examined: Corned beef (5 brands), ox cheek, ox tongue (3 kinds), collared head, tripe, oysters, sardines in oil, salmon, salmon cutlets, lobster, shrimps, curried fowl (2 kinds), boiled rabbit, boiled mutton, roast chicken, roast turkey, ox cheek soup, gravy soup, sausages, condensed milk (3 brands).

With the exception of the sausages, the whole of these samples contained more or less tin. One soup contained 35 mg of tin in a 1-pound can. A can of condensed milk contained 8 mg, and a can of preserved oysters 45 mg, besides a considerable quantity of copper. The metal is to be found throughout the mass of liquid soups and pasty curries, but resides chiefly on the outer surface of hard meats, such as corned beef. In many cases the canisters were much discolored and blackened on the inner surface, but in others the surface of the metal was perfectly bright, although there was an abundance of tin in solution. From the results given it appears beyond doubt that tin is readily acted upon by articles of food, vegetable and animal. Vegetable acids dissolve it abundantly, even if the contact is only of very short duration. Several samples of ginger ale and lemonade which were tested gave distinct tin reactions. Even CO_2 attacks the metal. Metallic tin readily precipitates lead from its solution, and from tin containing lead acids do not extract the lead until much of the tin has dissolved and the proportion of lead in the residue has become considerable. From solder lead can be dissolved simultaneously with the tin. It follows that there is little danger to be apprehended from the employment of impure tin for the manufacture of tin plates, the tin effectually preventing the lead from being dissolved. It is from the solder that contamination with lead might ensue. Regarding the physiological effects of tin salts, the following experiments were made: A half grown, apparently healthy guinea pig was given, together with its ordinary food, 25 mg of tin in the form of stannous hydrate. This had been freshly precipitated and had not been dried, but was given shaken up with water. There was no apparent effect. The solid excreta contained much tin after a few hours, while the metal could not be detected in the urine. Two days afterward the animal took 50 mg of tin as stannous hydrate. After three hours it appeared ill, and next morning it was dead. The quantity of feces passed since the administration of the second dose was very small, and the size of the feces had diminished extremely, about to that of those of a mouse. On dissection the stomach was found practically empty, the colon distended with food, the small intestines empty. The liver, kidneys, lung, and heart were separately examined. They all contained traces of tin, the largest quantity being apparently in the liver. The main part of the dose given, however, remained in the food contained in the colon, so that com-

---

[1] O. Hehner, Analyst, 1880, 5, 218.

paratively little of the oxid had been dissolved and absorbed. Death, therefore, had been produced by a far smaller quantity than that administered, and was apparently due to the astringent and irritant action of the metal.

Another somewhat stronger guinea pig took 30 mg of stannic hydrate, also freshly precipitated and moist. As no ill effect seemed produced, a further quantity of 45 mg was given on the same day. The feces contained much tin. Next day the animal took two further doses of stannic hydrate of 75 mg each. A few hours afterwards it appeared ill; its abdomen was distended, and the feces were diminishing in size. Next day the pig seemed quite well again, and took, without apparent effect, three doses of 75 mg each. Thus altogether it had in three days 450 mg of tin as stannic hydrate, without much injury, although the astringent effect of the tin had become visible.

On the following day, when it seemed in perfect health, it took 50 mg of tin in the form of stannous hydrate. It was ill next day and did not take any food until its death, three days afterward. The few excrements passed during that time were very small, much like those observed in the case of pig No. 1. They contained much tin. The stomach was practically empty; the colon and bowels filled with a semifluid, green, offensive matter, containing much tin. The liver contained a notable quantity of tin, and the lungs, heart, and kidneys traces of the metal.

From these experiments it appears that whilst stannic hydrate, from its comparative insolubility in gastric juice, is without much effect in the doses given, stannouhydrate, very soluble as it is in dilute acids, is a powerful irritant poison.

In a discussion of this paper before the Society of Analysts, Dr. Wynter Blyth suggested that the tin found by Mr. Hehner might not have been in a soluble form, but that minute particles of metallic tin might have been rubbed off by the mere friction of the contents of the can. He had administered 350 mg of finely divided tin at a dose and had seen no deleterious actions arise from it.

Mr. Wigner said that he could detect 20 or 30 mg of tin per pound by taste. He had only found one sample of canned fish which was free from tin present in the fish. This was a can of prawns and they had probably been canned less than a month. He did not think that in the case of fish it was, as Dr. Blyth had suggested, mechanical adherence of the tin. In nearly every case of condensed milk which had been kept more than one month or six weeks, tin and lead were both present. He had lately examined 50 brands of tongues, hams, chicken, corned beef, and roast beef, and had found only one can in which tin was present in any appreciable quantity. There was about $\frac{1}{4}$ mg in a pound. He believed that the solder used often contained bismuth. He had also examined more than 300 kinds of canned fruits and only 8 or 10 had turned out bad.

USE OF TIN CANS FOR PRESERVING.[2]

Niederstadt calls attention to the fact that acid-preserved foods dissolve tin. Green foods (beans, peas, pickles, etc.) contain salts which form double salts with tin compounds. Preservation with SO_2 should not be allowed on account of the solubility of tin in sulphuric and sulphurous acid.

POISONOUS ACTION OF TIN.[2]

In conclusion to investigations made by the authors some time since[3] upon tin in canned foods, they have made experiments upon the toxicity of tin combinations. That the tin which is present in these foods in a difficulty soluble form was dissolved

---

[1] Niederstadt, Apoth. Ztg. 1891, 6, 588; abs. Chem. Centrbl., 1892, 1, 62.
[2] Emil Ungar and Guido Bodländer, Ztsch. Hyg., 1887, 2, 241; abs. Chem. Centrbl., 1887, 644.
[3] Chem. Centrbl., 1883, 810.

and absorbed during digestion was proved by the detection of tin in different organs of a dog and a rabbit fed with food containing tin. Subsequently in two cases urinary analysis showed that with men also a part of the tin introduced into the stomach by such foods was absorbed. Experiments were made to settle the question how far the introduction of tin into the circulation could work injury. Those which were made on three species of animals—frogs, rabbits, and dogs—with subcutaneous injections of sodium-stannous tartrate showed that the introduction of a noncaustic tin salt into the animal organism, even though it be not introduced directly into the circulation, caused pathological symptoms and finally death.

The conclusion was drawn that even minimal doses of tin, when frequently repeated, destroy health and finally cause death.

With regard to the possibility of chronic tin poisoning it was established that small doses of tintriethyl acetate subcutaneously injected, when frequently given, caused an intoxication culminating in death, thus corroborating the results of White.[1] It is found that the toxic action of tintriethyl acetate is much greater than that of sodium-stannous tartrate, and it appears that the poisonous action of this compound is not due to the metal alone. Tintriethyl acetate, sodium-stannous tartrate, and stannous chlorid (in milk) were also introduced through the mouth. These experiments showed the poisonous qualities of tin combinations. The authors believe they are able to answer in the affirmative the question whether food containing tin can cause a general intoxication or a chronic tin poisoning, aside from an accidental local action.

TECHNICAL DETERMINATION OF ZINC.[2]

Prepare a solution of zinc ferrocyanid by dissolving 44 grams of the pure salt in water and diluting to a liter. Standardize as follows:

Dissolve exactly 300 mg of pure zinc oxid in a beaker in 10 cc of strong hydrochloric acid. Now add 7 grams of ammonium chlorid and about 100 cc. of boiling water. Titrate the clear liquid with the ferrocyanid solution until a drop, when tested on a porcelain plate with a strong aqueous solution of uranium acetate, shows a brown tinge. About 16 cc of ferrocyanid will be required, and accordingly nearly this amount may be run in rapidly before making a test, and then the titration finished carefully by testing after each additional drop of ferrocyanid. As soon as a brown tinge is obtained, note the reading of the burette, and then wait a minute or two and observe if one or more of the previous tests do not also develop a brown tinge. Usually the end point will be found to have been passed by a test or two, and the proper correction must then be applied to the burette reading. Finally make a further deduction from the burette reading of the amount of ferrocyanid required to produce a brown tinge under the same conditions when no zinc is present. This correction is about two drops or 0.14 cc.

Two hundred mg of zinc oxid contain 160.4 mg of zinc and one cc of the above standard solution will equal about 0.01 gram of zinc or about 1 per cent where one gram of ore is taken for assay.

Prepare the following for the assay of ores:

(1) A saturated solution of potassium chlorate in nitric acid, made by shaking an excess of the crystals with the strong, pure acid in a flask. Keep the solution in an open flask.

(2) A dilute solution of ammonium chlorid containing about 10 grams to the liter. For use heat to boiling in a wash bottle.

(3) A wash bottle of hot water:

Take exactly one gram of ore and treat in a 3.5 inch casserole with 25 cc of the chlorate solution. Do not cover the casserole at first, but warm gently until any

---
[1] Arch. exp. Pathol. u. Pharm., 13, 53.
[2] Von Schulz and Low. J. Analytical and applied Chem., 1892, 6, 491.

violent action is over, and greenish vapors have ceased to come off. Then cover with a watch glass and boil rapidly to complete dryness, but avoid overheating and baking. A drop of nitric acid adhering to the cover does no harm. Cool sufficiently and add 7 grams of ammonium chlorid, 15 cc of strong ammonia and 25 cc of hot water. Boil the covered mixture one minute, and then, with a rubber-tipped glass rod, see that all solid matter on the cover, sides and bottom is either dissolved or disintegrated. Filter into a beaker and wash several times with the hot ammonium chlorid solution. A blue-colored filtrate indicates copper. In that case add 25 cc of strong pure hydrochloric acid, and about 40 grams of granulated test lead. Stir the lead about in the beaker till the liquid has become perfectly colorless, and then a little longer to make sure that the copper is all precipitated. The solution, which should be still quite hot, is now ready for titration. In the absence of copper the lead is omitted and only the acid added. About one-third of the solution is now set aside and the main portion is titrated rapidly with the ferrocyanid till the end point is passed, using the uranium indicator as in the standardization. The greater portion of the reserve portion is now added, and the titration continued with more caution till the end point is again passed. Then add the remainder of the reserved portion and finish the titration carefully, ordinarily by additions of two drops of ferrocyanid at a time. Make corrections of the final reading of the burette precisely as in the standardization.

Gold, silver, lead, copper, iron, manganese, and the ordinary constituents of ores, do not interfere with the above scheme. Cadmium behaves like zinc.

DETECTION OF BENZOIC ACID.[1]

Extract the substance with ether, evaporate off the ether from the resulting extract, and treat the residue with two or three cc of strong sulphuric acid. Heat till white fumes appear. Organic matter is charred and benzoic acid is converted into sulphobenzoic acid. A few crystals of potassium nitrate are then added. This causes the formation of metadinitrobenzoic acid. When cool the acid is poured into water and ammonia added in excess, followed by a drop or two of ammonium sulphid. The nitro compound becomes converted into ammonium metadiamidobenzoate, which possesses a peculiar reddish brown color. The benzoic acid must first be separated in a state of approximate purity before this test can be applied. Half a milligram of the acid can be detected. in the absence of interfering bodies.

---

[1] Mohler, Bull. soc. chim., (3), **3**, 414.

INDEX.

A.

	Page
Accum, food adulteration	1073
Ader, M., & Cie	1087, 1146
Adulteration of canned foods	1025
Air, influence in producing decay	1023
Albuminoids, determination of	1028
digestible, determination of	1028
Almonds, copper in	1067
Altemus, F. E	1085, 1121
American peas, copper in	1017, 1076
Amherst Packing Co	1108, 1115, 1146
Amieux Frères	1085, 1088, 1106, 1140, 1146
Anderson, A	1132, 1147
Anderson Preserving Co	1129, 1147
Angilbert, P. A	1022
Appert	1023
Apricots, copper in	1067
Artichokes	1127
Ash, determination of	1029
corrected	1029
Asparagus	1134
Atlantic Canning Co	1124, 1147
Aubin-Salles, Vve	1084, 1147
Aughinbaugh Canning Co	1112, 1147
Austin, Nichols & Co	1121, 1147

B.

Bacon, P. F	1118
Bacteria, influence of heat on	1015, 1023
salicylic acid, etc., on	1030
Baile & Stouffer	1121, 1142, 1147
Baked beans	1113
Baker, J. C	1119, 1147
Baker & Brown	1109, 1147
Bamberger, T. W., & Co	1101, 1147
Bamberger & Brewington	1101, 1147
Barbour, J. L., & Son	1120
Baron, Père et Fils	1088, 1147
Barth, remarks on copper-greening	1068
Barton Fils	1160, 1161
Basset & Fogg	1132, 1147
Batavia Preserving Co	1092, 1112, 1117, 1120, 1137, 1147

	Page.
Bavarian report (1891) on copper-greening	1067
(1892)	1069
Beall & Baker	1083, 1084, 1100, 1118
Beans, baked	1113
haricots flageolets	1106
panachés	1107
vertsc	1097
Lima	1109
little green	1108
red kidney	1117
string	1020, 1099
stringless	1102
wax	1109
Beebe's method for hydronaphthol	1031
Belgium, copper-greening in	1066
Benzoic acid	1035, 1055
detection of	1032, 1167
Bergeron, Bussy and Fauvel, report on copper-greening	1048
Bernard, Alexandre	1160, 1161
Berry, E. E.	1084, 1111, 1121, 1142
Bevan	1024
Biardot, etc., calcium-sucrate method of greening	1051
Billet	1160, 1161
Birch & Co.	1085, 1111, 1122
Blythe, A. W.	1041, 1165
Blair Canning Co	1124, 1147
Bodländer, Guido and Emil Ungar	1042, 1165
Bonney, Wheeler, Dingley & Co	1124, 1148
Booth, A., Packing Co	1110, 1148
Borax	1035
Borgmann	1071
Boric acid	1035
Bolland & Grosset	1160
Bouchardat and Gautier	1064
report on copper-greening	1050
Bouley	1057
Bouvais	1160, 1161
Bowen, B. F.	1077, 1078
Boyle, Granville H	1121, 1148
Brard & Cocary	1090, 1148
Brass vessels, use of	1018
Bread, copper in	1044
Brennard Frères	1160
Brewington, Chas., & Co	1079, 1148
Briant	1160, 1161
Bright plate	1035
Broadmeadow, J., & Son	1135, 1148
Brooklyn board of health	1025, 1074, 1150, 1159
Brooklyn, prohibition of the sale of coppered pickles in	1159
Brouardel	1047, 1057
Pasteur, etc., report on copper-greening	1055
Riche, etc., report on copper-greening	1054
Wurtz, etc., report on copper-greening	1056
Brown, A. F.	1120, 1148
Brown, Jas. K., & Co	1120, 1162

INDEX. III

	Page.
Browning & Middleton	1079, 1080, 1081, 1100, 1111, 1115, 1118, 1120, 1129, 1134, 1140
Brunswick copper-greening case	1068, 1072
Brussels sprouts	1131
Bryan, J. B., & Bro	1079, 1083, 1084, 1097, 1099, 1104, 1110, 1114, 1129, 1134, 1140, 1145
Bubb, F. L.	1111, 1142, 1144
Bucklin, C. S	1138, 1148
Bull, Jacob E	1133, 1148
Bunting, T. L	1077
Burchard, G. C	1079, 1083, 1119, 1142
Burnham, A. H	1121, 1142, 1148
Burnham & Morrill	1115, 1143, 1148
Bussy	1052
Bussy, Fauvel, etc., report on copper-greening	1048
Ville, etc., report on copper-greening	1046
Wurtz, etc., report on chlorophyl-greening	1049, 1050, 1149

C.

Cadeau, Geo., & Cie	1088, 1148, 1160
Calbiac, René	1135, 1136, 1148
Cahours	1033
Calcium-sucrate method of greening	1051
Canning food, old methods	1022
Cannon, H. P	1090, 1148
Cans, quality of tin plate used for	1018, 1036
Capers, copper in	1161
Carles	1047, 1052
Cicero Canning Co.	1091, 1148
Charpentier	1160, 1161
Chatin	1060
Chatin, Brouardel, etc., report on copper-greening	1056
Cherries, copper in	1067
Chevallier, A	1047
Chinois, copper in	1067
Chlorophyl, action of cooking on	1043
copper salts on	1043, 1070
greening process	1049, 1051, 1056
Church, A. H., copper in feathers of turaco	1043
Cirio, Francesco	1083, 1149
Clagett, T	1121, 1149
Clark, John	1074
Clark, T. W., & Son	1080, 1149, 1151
Clot, J., & Cie	1043, 1069, 1162
Cobb, A. H	1142
Constant Fils	1106
Copper, determination of	1039, 1052, 1069, 1163
Coppered peas, digestibility of	1162
Copper-greened vegetables, recognition of	1048
Copper-greening, electrolytic	1043, 1069, 1162
French prohibition of	1046
in Belgium	1066
France	1046
Germany	1066
Great Britain	1073
Italy	1072

	Page.
Copper, normal occurrence in feathers	1043
food	1043
phyllocyanate	1043, 1070
physiological action of	1045, 1051, 1054, 1056, 1060, 1068, 1070
vessels, use of	1016, 1042, 1046, 1058
Corn	1118
Corn sweetener	1054
Corn and tomatoes, mixed	1144
Cornwell, G. G., & Son	1098, 1128, 1133, 1137, 1139
Corrected ash	1029
Cost of food as found in canned vegetables	1020, 1021
Couteau	1085, 1149, 1160
Couteaux, C	1083, 1149
Cox, J. T	1079, 1149
Crown, M. F	1086, 1101, 1122
Cucumbers, copper in	1067
Currants, copper in	1067
Currie, Donald	1024
Curtice Brothers Co	1001, 1104, 1114, 1116, 1124, 1137, 1139, 1149

D.

Dadelzen, E. M	1160, 1161
Daly, J. J	1081, 1110, 1114, 1120, 1132, 1133
Dandicolle & Gaudin	1038, 1084, 1086, 1087, 1097, 1108, 1131, 1134, 1135, 1136, 1149, 1160, 1161
Davis, Baxter & Co	1123, 1150
Davis, I. T., meat preservatives	1030
Davis, William T	1084, 1101
Decaisne	1057
De Haen	1069
De Heine, Aug	1024
Digestible albuminoids, determination of	1028
Digestibility of canned vegetables	1021
coppered vegetables	1065, 1162
Digestion, artificial, Niebling's method	1028
Dry matter, amount found in canned vegetables	1020
Dufour, E., & Cie	1160, 1161
Dunbar's Sons, G. W	1128, 1130, 1150
Dumagnou	1048
Dupin, A	1160, 1161
Dupont, Jules	1083, 1150, 1160
Duprat, Clément, & Maurel	1160, 1161
Dupré, A., and Odling, copper in food	1044
Durand, Pierre	1160, 1161
Du Raix, Eugène	1038, 1086, 1098, 1106, 1107, 1135, 1150, 1160, 1161

E.

Earnshaw, J. T	1079, 1082
Egger	1068
Electrolytic copper-greening	1043, 1069, 1162
Erie Preserving Company	1122, 1150
Estler Bros. & Co	1080, 1083, 1100, 1115
Ether extract	1028

INDEX. V

	Page.
Evans, Day & Co.	1110, 1111, 1150
Ewald	1034
Excelsior Canning Co	1100, 1150
Eyquem, A	1140, 1150, 1160, 1161

F.

Fait & Slaglo	1133, 1150
Fait & Winebrenner	1080, 1129, 1133, 1142, 1149, 1150, 1156
Feathers, copper in	1043
Farren, J. S	1100, 1151
Fenton, C. M	1122
Fernand, Eugène & Cie	1160
Fiber, crude	1028
Figs, copper in	1067
Fiton, J., Aîné & Cie	1089, 1151, 1160, 1161
Flour, copper in	1044
Fontaine Frères	1098, 1128, 1151, 1160, 1161
Food preservatives	1016
Food value of canned vegetables	1020
Foos, C	1144, 1151
Forestville Canning Co	1123, 1137, 1151
France, copper-greening in	1046
Franklinville Canning Co	1079, 1151
Frederick City Packing Co	1143, 1151
Fremont Canning Co	1132, 1151
Fremy	1037
French peas, copper in	1017, 1075
French string beans	1097
Fries, Alex., & ''os	1035
Fyfe, Peter	1074

G.

Galippe	1051, 1052, 1054, 1060, 1072
physiological effects of copper salts	1045, 1048, 1058
report on copper-greening	1057
use of copper cooking utensils	1058
Gallard	1063, 1064
report on copper-greening	1059
Wurtz, etc., report on copper-greening	1056
Gardner, W. L	1104, 1151
Garges	1050, 1051
Garres, Vve., & Fils	1041, 1078, 1087, 1151, 1160, 1161
Gautier	1054, 1057, 1058
and Bouchardat, report on copper-greening	1050
General examination of canned vegetables	1027
Gerland	1033
German law relative to tin plate	1018, 1019, 1036
solder	1019
Germany, copper-greening in	1066
Gibbs Preserving Co	1081, 1151
Gillard	1160, 1161
Girard, Gallard, etc., report on copper-greening	1056
Githens, Rexsamer & Co	1089, 1130, 1152

	Page.
Githens & Rexsamer	1101, 1104, 1122, 1142, 1145, 1151
Glasgow, city of, report on copper-greening	1073
Glenwood Canning Co	1132, 1152
Gley	1064
Glycerophosphate of copper	1058, 1072
Gobelin, Fils & Cie	1084, 1085, 1152
Godillot, jne., A	1098, 1152, 1160
Gooseberries, copper in	1067
Gordon & Dilworth	1145, 1152
Gouleau	1160, 1161
Great Britain, copper-greening in	1073
Green gages, copper in	1067
Grimaux, Ed	1063, 1065, 1066, 1075, 1076, 1084
report on copper-greening	1064
Grocers' Packing Co	1114, 1115, 1152
Guillaumez	1088, 1089, 1152, 1160, 1161
Guillemare and Lecourt	1051
petition of	1049

H.

Hagan & Gilkison	1085, 1115
Halenke	1071, 1072
Hamburgh Canning Co	1077, 1152
Hardell, John W	1086, 1101
Haricots flageolets	1106
panachés	1107
verts	1097
Harrison, Chas. K	1123, 1152
Hassall, A. H	1073
Hazelnuts, copper in,	1067
Heat, influence of, on bacteria	1015, 1029
Hehner, O	1035, 1164
tin in canned goods	1042
Hemingway, H. F., & Co	1091, 1152
Hilger	1071, 1072
Historical	1022
Hörmann	1072
Houston, I. H	1118, 1132, 1152
Hudson & Co	1135, 1152
Hume, Frank	1078, 1082, 1083, 1097, 1100, 1118, 1119, 1121, 1140
Hungerford, J. H	1115, 1123, 1087, 1098
Hunt, G. W., & Co	1082, 1153
Hydronaphthol	1035
detection of	1031

I.

Italy, copper-greening in	1072

J.

Jackson & Co	1079, 1080, 1084, 1110, 1114, 1118, 1119
James, W. L	1123, 1153
Jenkins, T. M	1058
Jones and Trevithick	1024
Julien, Chs	1160, 1161

K.

	Page.
Kämmerer	1071
Karsch	1067
Kayser	1068, 1071
Keagle & Guider	1090, 1153
Kellogg, Chas. I	1090, 1098, 1104, 1108, 1136, 1143
Kengla, H	1123
Kennedy, G. E., & Co	1091, 1102, 1106, 1135, 1140
Kent, W. H	1066, 1074
V. Kerschensteiner	1071
Keyworth, John	1088, 1111, 1135
Kidney beans, red	1044, 1069
Kidneys, copper in	1117
Koehler	1069
Kolbe, H	1033, 1034

L.

Laing, Charles, & Co	1087, 1153
Lamau, François, & Co	1160, 1161
Lambert, Henri, & Cie	1086, 1090, 1153, 1160, 1161
Lancet commission	1073
Lanfair, H. S., & Co	1100, 1153
Laurel Canning Co	1137, 1153
Lautemann	1033
Lead, action on tin solutions	1035, 1163
determination of	1039, 1052
in canned goods	1039
Lead-poisoning	1041
Lead in rubber rings	1019, 1102, 1163
Lead in solder	1037
Lead in tin plate	1035, 1058, 1162
German law relative to	1018, 1019, 1036
Lead, physiological effects of	1018, 1040
Lead-tin alloys, action of food on	1058, 1163
Lead-topped bottles	1019, 1038
Lecourt and Guillemare	1049, 1051
Leggett, F. H., & Co	1077, 1090, 1110, 1119, 1129, 1134, 1139, 1145, 1153
Lehmann	1034, 1071
report on copper-greening	1069
Leignette	1024
Le Lagadec	1086, 1153
Lemouies	1160, 1161
Leopold, V	1085, 1153, 1160, 1161
Lewis, C., & Co	1114, 1153
Lewis, W. K	1114
Lima beans	1109
List of packers	1146, 1160
Liver, copper in	1044, 1069
Los Angeles Packing Co	1111, 1153
Love, John P	1086, 1098, 1106, 1122, 1123, 1142, 1145
Louit Frères & Co	1098, 1154
Low and Von Schulz	1166
Lowekamp, J. F	1078, 1100, 1154
Ludington, J., & Co	1082, 1087, 1154

M.

	Page.
Macédoino	1139, 1162
proximate composition of	1140
McGaw, C. A	1119, 1154
McGrath, H. J., & Co	1101, 1154
McMurray, Louis, Packing Co	1111, 1154
McShane, John	1144, 1154
McWaid & Martin	1124, 1147
Mach	1069
Magruder, J. H	1136
Maine State Packing Co	1111, 1122, 1154
Mallory, E. B., & Co	1101, 1154
Marcelino	1160, 1161
Marcies & Cie	1160, 1161
Markell Bros	1090, 1154
Martin, A. J	1057
Martin, Miss Maggie	1074
Massachusetts board of health	1025, 1149, 1153, 1159
sale of coppered vegetables in	1074
Mattocks, Charles P	1110, 1119, 1154
Mayrhofer	1069, 1071, 1072
report on copper-greening	1067
Meader, H. I	1079, 1121
Medlars, copper in	1067
Mercier, Eugène	1160, 1161
Merkel, G	1068
Merrell & Soule	1111, 1142, 1154
Metals, heavy, determination of	1039, 1052
Metal vessels for culinary purposes	1016, 1042, 1046, 1058, 1163
Meyer, T. J., & Co	1142, 1154
Micé	1047
Mick, Bernard	1124
Micro organisms, destruction of	1015, 1023, 1029
Miller Brothers & Co	1077, 1154
Minium in rubber rings	1163
Mitchell Bros	1118, 1155
Mohler, method for benzoic acid	1167
Moir, John, & Son	1106, 1155
Monbadou, R	1160, 1161
Montgomery, C. F	1115
Mound City Preserving Co	1091, 1105, 1155
Mushrooms, copper in	1161
Muter, J	1162
Myer, Thos. J., & Co	1081, 1099, 1123, 1155

N.

Nadal, Amédée	1082, 1083, 1084, 1140, 1155
Nail City Packing Co	1090
Napias	1057
Nebraska City Canning Co	1124, 1155
Nelson, C. E	1078, 1120
Nelson, O	1091, 1105, 1124, 1132
New Jersey dairy commission	1074, 1161
Nicholas & Cie	1160, 1161

INDEX.

	Page.
Nicolas, S., & Cie	1088, 1155
Niebling, method for artificial digestion	1028
Niederstadt, E	1165
Nitrogen, determination of	1028
Northern Maine Packing Co	1123, 1155
Nouvialle, J., & Cie	1088, 1128, 1155
Nouville, J., & Cie	1160, 1161
Numsen, Wm., & Sons	1078, 1080, 1099, 1110, 1155
Nunley, Hynes & Co	1080, 1120, 1151, 1156

O.

Ogier, digestibility of coppered vegetables	1065
Okra	1129
Okra and tomatoes, mixed	1145
Oxygen, influence in producing decay	1023
Oysters, copper in	1069

P.

Packers, list of	1146, 1160, 1161
Page, J. F.	1090, 1101, 1137, 1139, 1142
Pancreas solution	1028
Parisian Canners' Association	1048, 1063
Pasch	1072
Parson Bros	1078, 1156
Parsons, A. M.	1091, 1092, 1112, 1116, 1117, 1129
Pasteur	1052, 1054, 1060
Pasteur, Poggiale, and Brouardel, report on copper-greening	1055
Pasteur, report on copper-greening	1047
Paul, B. H., and C. T. Kingzett, copper in food	1162
Peaches, copper in	1067
Peas, analytical data	1094
calculate to dry substance	1096
copper in	1017, 1075
description of samples examined	1075
weights of samples	1092
Pears, copper in	1067
Pearson, C. H., Packing Co	1102, 1156
Perron	1037
Personne	1060
Phyllocyanate of copper	1043, 1070
Phyllocyanic acid	1043
Pickles, coppered	1074
prohibition of sale in Brooklyn	1159
Piesse, Charles H., analysis of canned peas	1163
Pinard, Alphonse	1160, 1161
Pinette, J	1036
Piria	1033
Poggiale, Pasteur, and Brouardel, report on copper-greening	1055
Possoz, etc., calcium-sucrate method of greening	1051
Potato, sweet	1129
Potter & Wrightington	1113, 1114, 1115, 1152
Preservatives	1016, 1029
methods for the detection of	1030
Preservatives, use of	1025

	Page.
Preservative, use of common salt as a	1021
Preserving, methods of	1015, 1022
Price, L. A	1035, 1091, 1140, 1156, 1161
Proctor, C. W	1099, 1110, 1113, 1118, 1132
Proctor, salicylic acid	1033
Proust	1057
Ptomaines, occurrence of in canned vegetables	1025
Pumpkins	1137

R.

Ranso	1058
Raspberries, copper in	1067
Raymond, J. H., sale of coppered pickles	1159
Rettie	1024
Reuss, W., rubber rings containing lead	1163
Richardson & Robbins	1135, 1156
Riche	1060
Riche, Brouardel, etc., report on copper-greening	1054
Rilhac	1100, 1161
Risch & Cheminant	1140, 1156
Roberts, Thos., & Co	1141, 1156
Robinet & Cie	1085, 1156
Rochard	1057
Rodel & Fils	1100, 1161
Roe, A. B	1080, 1156
Root, John, & Sons	1118, 1122, 1156
Roudenct, F	1160, 1161
Royal Preserving Co	1110, 1156
Rubber rings, lead in	1019, 1102, 1163
Russell, James D	1074
Russell, J. F	1088, 1089, 1104, 1123, 1142, 1145
Ryan	1024

S.

Saccharin	1034
detection of	1032
Saddington	1022
Salt	1021
determination of	1029
Salicylic acid	1055
alleged case of poisoning by	1026
detection of	1031, 1034
in beer	1033
physiological effects	1.33
use as a food preservative	1026
Samples, buying of	1022, 1074
Satterfield, W. C	1086, 1157
Schoffer	1058
Schuster & Knox	1091, 1112, 1132
Schwinghammer, John	1133, 1157
Scudder & Townsend	1134
Sendtner, R	1072
Shea, N. H	1080, 1081, 1082, 1100, 1110, 1114, 1118, 1141
Sherwood, J. R	1121

INDEX. XI

	Page.
Shriver, B. F., & Co	1079, 1080, 1100, 1101, 1104, 1118, 1120, 1121, 1157
Sisk, A. W	1087, 1157
Smith, Yingling & Co	1121, 1157
Solder	1037
analyses	1038
German law relative to	1019
in canned foods	1040
presence in canned foods	1039, 1052
separation of	1027
use in canning	1037
Soulé & Price	1160, 1161
Source, Maguier de la	1060
Brouardel, etc., report on copper-greening	1054
Squash	1139
Steele Brothers	1102, 1157
Stevens, W. L	1132, 1157
Strawberries, copper in	1067
String beans	1099
Stringless beans	1102
Stutzer, pancreas solution	1028
Succotash	1141
Sulphurous acid	1032
detection of	1031, 1032
Summers, C. G., & Co	1120, 1157
Sweet potato	1129
Sympson	1162

T.

	Page.
Talbot Frères	1078, 1097, 1157, 1160, 1161
Tardieu, Bussy, etc., report on copper-greening	1046
Taste of vegetables greened with copper	1053, 1061, 1065, 1069
Tatlock, R. R	1074
Taylor, H. E	1035
Terne plate	1035
Tertrais, Victor	1160, 1161
Thurber, Whyland & Co	1104, 1111, 1122, 1135, 1157
Tin, action of foods on	1164
on lead solutions	1035, 1163
German law relative to	1018, 1036
in canned goods	1041
determination of	1039, 1052, 1164
lead alloys, action of food on	1058
occurrence in food	1164
physiological action of	1042, 1164
plate	1035
plate, lead in	1018
sulphid	1042
Tirard, circular relative to copper-greening	1059
Tisserand et Fils	1084, 1158, 1160
Tomatoes	1131
Tomatoes and corn, mixed	1144
okra, mixed	1145
Toussaint	1051
Towle & Morian	1091, 1102, 1112, 1124, 1132

	Page.
Trélat, E	1057
Trevithick and Jones	1024
Triat, Gabriel, & Co	1083, 1158, 1160, 1161
Tschirch	1070, 1071
action of copper in greening process	1043
Tucker, S. S	1081, 1101, 1110, 1119, 1120, 1133
Turaco, copper in the feathers of	1044
Turnip, copper in	1044

U.

Ungar, E., and G. Bodländer, physiological action of tin	1042, 1165

V.

Van Camp Packing Co	1081, 1158
Ville, Bussy, etc., report on copper-greening	1046
Von Schulz & Low	1166

W.

Wagner, Martin, Co	1079, 1158
Wallace, Jas., & Son	1086, 1132, 1158
Water in canned foods	1020, 1021, 1027, 1028
Wax beans	1109
Wayne County Preserving Co	1115, 1142, 1158
Western New York Preserving & M'f'g Co	1081, 1120, 1158
Wheat, copper in	1014
White, Robert	1087, 1135, 1144, 1166
Whiteford, W. S	1082, 1158
Wigner, G. W., copper in food	1163
tin in food	1165
Wiley, H. W., introduction and summary	1015
Williamson, R., & Co	1119, 1158
Winebrenner, P. F. & D. E	1099, 1100, 1158
Winfield, A. A	1078, 1079, 1100, 1119, 1120, 1132, 1137
Wolffhügel, Gustav, lead in tin plate	1035, 1163
Wright, J. A., & Bro	1100, 1106, 1123, 1158
Wurtz	1060
Wurtz and Bussy, report on copper-greening	1149
Wurtz, Gallard, etc., report on copper-greening	1056

Y.

Yewell, E. L	1104
Youngs, Elphonzo, Co	1088, 1089, 1090, 1098, 1104, 1107, 1111, 1130, 1135, 1138, 1143

Z.

Zinc	1041
determination of	1039, 1052, 1166
physiological effect of	1041
process for greening vegetables	1016, 1051

www.ingramcontent.com/pod-product-compliance
Lightning Source LLC
Chambersburg PA
CBHW020304170426
43202CB00008B/492